The Apollo Story

This remarkable photograph showing a connecting ring during separation of the Saturn V's first and second stage, was captured by an automatic camera on the unmanned Apollo 6 in April 1968. The exposed film was ejected and parachuted back to Earth in a protective capsule.

Title page: *Harrison Schmitt of Apollo 17 collecting samples from beside a split bolder.*

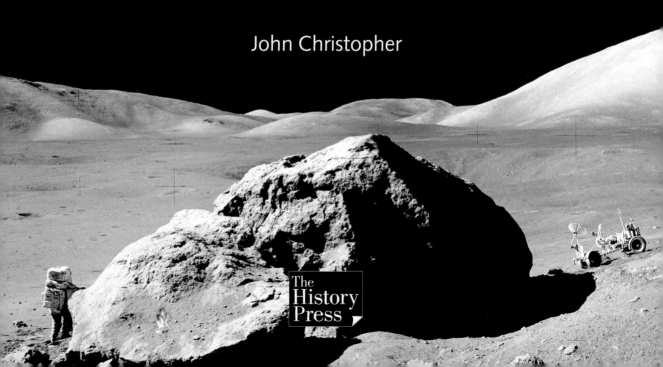

The Apollo Story

John Christopher

The History Press

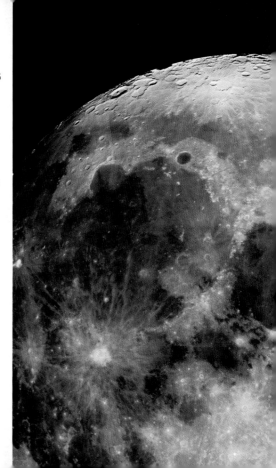

Published in the United Kingdom in 2009 by
The History Press
The Mill · Brimscombe Port · Stroud · Gloucestershire · GL5 2QG

Copyright © John Christopher, 2009

British Library Cataloguing in Publication Data
A catalogue record for this book is available from the British Library.

Hardback ISBN 978-0-7524-5173-2

Typesetting and origination by The History Press
Printed in Italy

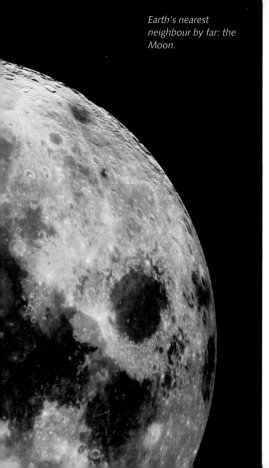

Earth's nearest neighbour by far: the Moon.

CONTENTS

ACKNOWLEDGEMENTS

In my experience the creation of any book such as this involves assistance from a number of sources, but in this instance I have only one to thank: all of the images in *The Apollo Story* are from the archives of NASA.

In addition I need to thank them for more than that, as I am a child of the space age, born fourteen months before the plaintive bleeps of Sputnik heralded the start of the space race. I was four years old when Yuri Gagarin went into orbit and not quite thirteen by the time Neil Armstrong and Buzz Aldrin stepped onto the Moon. Consequently, the Sixties were an incredibly exciting time to be growing up in. For a start, we had a new type of science fiction to feed the imagination with *Dr Who*, Gerry Anderson's *Thunderbirds* and *Captain Scarlet*. There was even talk of something called *Star Trek* coming over from the USA. But far, far better than all of this was the fantastic real-life adventure unfolding before our eyes. And while other boys collected cards of their favourite football players, I could recite the name of each and every astronaut and even their Soviet counterparts, the cosmonauts.

In the short span of only eight years or so, manned space exploration had gone from zero to the Moon. The Apollo programme burst across the heavens like a spectacular firework, then faded away almost as quickly. I followed the events described in this book at the best of times, and I think that sense of adventure and achievement has stayed with me ever since. In no small part Apollo pointed me on my course through life; I now fly as a professional balloon pilot as well as working with airships. Through ballooning I had the good fortune to become involved in a second great adventure in the 1990s as several teams vied to be the first to fly non-stop around the world. While I was in Morocco with Richard Branson's Virgin Global Challenger team, I happened to have breakfast with one of my childhood heroes. His name is Buzz Aldrin, and that's him on the right.

So, thank you NASA... for everything.

Strictly speaking, that's not the only acknowledgement. For her continued support and proof-reading diligence I am also indebted to my wife Ute, and for Anna and Jay I hope this account of Apollo will serve to convey some of the magic and excitement of my own childhood.

John Christopher

Neil Armstrong and the lunar module of Apollo 11 are reflected in Buzz Aldrin's spacesuit visor, which has been coated with a thin layer of gold to protect the astronaut from the harsh glare of the Sun.

PRELUDE TO APOLLO

"I believe that this nation should commit itself to achieving the goal, before the decade is out, of landing a man on the Moon and safely returning him to Earth. No single space project in this period will be more impressive to mankind, or more important for the long-range exploration of space; and none will be so difficult to accomplish."

President John F. Kennedy's address to a joint session of Congress on 25 May 1961 was a bold and unequivocal statement of intent; a gauntlet thrown down at a time when America was reeling from a series of spectacular space achievements by their Cold War adversaries, the Soviet Union. In April 1961 Yuri Gagarin became the first man into space when he orbited the Earth in Vostok 1, leaving Washington agog with

➤

Riding a white-hot column of raw energy from the Saturn V's massive first-stage engines, Apollo 15 is launched from Cape Kennedy, Florida, on the fourth lunar landing mission.

fears of a Soviet domination of space, a perceived 'missile-gap' and even talk of a red flag on the Moon.

On 5 May the Americans responded by launching Alan Shepard aboard Mercury 3 – by comparison with Vostok 1 a paltry

President Kennedy reiterates the goals and needs of the space programme during a speech at Rice University, Texas, in 1962. He said they were doing it not because it was easy, but because it was hard.

sub-orbital lob lasting only fifteen minutes. Clearly they were lagging a long way behind in the space race, and although the seeds of the Apollo programme had been sown in the last years of President Eisenhower's administration, it was Kennedy who recognised the quest for the Moon as a potent weapon against his enemies and a rallying point for the nation. As he later reiterated during a speech given at Rice University in Texas, 'We choose to go to the Moon in this decade, and do other things, not because they are easy, but because they are hard.'

Never a truer word was spoken. Reaching this desolate, air-less world approximately 240,000 miles (386,000km) from the Earth involved overcoming some formidable challenges. The task fell to NASA, the National Aeronautics and Space Administration, and initially there were four main mission modes under consideration:

Direct Ascent, with an immensely powerful booster sending a spacecraft to the Moon, landing and returning as a single unit.

Earth Orbit Rendezvous (EOR), with a number of rockets carrying component sections for assembly in Earth orbit before heading to the Moon.

Lunar Surface Rendezvous (LSR), which envisaged two spacecraft launched in succession, with one carrying propellants followed by a manned vehicle which would then refuel from the other for the return leg.

Lunar Orbit Rendezvous (LOR), with a single launch vehicle carrying a spacecraft of modular parts, with the command module remaining in orbit around the Moon while a specialised lander would descend to the surface and then return to dock in lunar orbit.

When Kennedy addressed Congress in 1961 the Direct Ascent mode was widely favoured by NASA, but with time the proponents of the Lunar Orbit Rendezvous won through and Apollo's final configuration began to take shape. The main advantage of the LOR scenario is that it required only a small lander to descend to the surface, thus minimising the mass to be launched back into orbit.

At the tip of the Apollo spacecraft is the command module (CM), a capsule

◄

In the direct ascent scenario, the capsule at the tip of the rocket stack blasts-off from the lunar surface on the return journey to Earth. Direct ascent might do away with the need for complicated docking manoeuvres, but on the downside it would require a launch vehicle even bigger than the Saturn V.

for three astronauts. Next comes the cylindrical service module (SM) housing the main engine, propellants, a fuel-cell power system, manoeuvring thrusters and consumables such as water and air. Attached together they form a unit known as the CSM, separated only for the CM's final re-entry and splashdown. Beneath this and partly folded within a protective shroud comes the lunar module (LM) – a two-stage craft to carry two astronauts down to the surface using the engine on

'We have seen a wonder. There has never been one quite like it. What first steps in human history would one have chosen to witness, if one could travel in time.'

Lord C.P. Snow, 1969

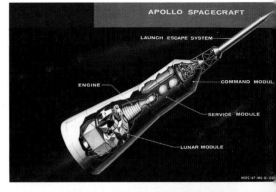

APOLLO SPACECRAFT

LAUNCH ESCAPE SYSTEM

ENGINE

COMMAND MODUL

SERVICE MODULE

LUNAR MODULE

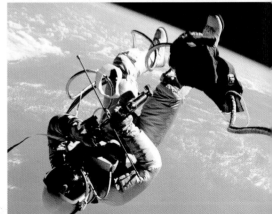

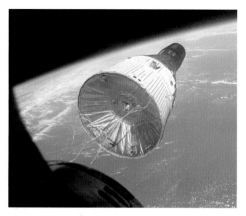

Did you know?

The only Mercury astronauts to fly on Apollo all had surnames beginning with the letter 'S'. Out of Wally Schirra (Apollo 7), Alan Shepard (Apollo 14) and Deke Slayton (ASTP), only Shepard went to the Moon.

◀◀

Close encounter: the rear of the Gemini 7 capsule photographed from Gemini 6 in December 1966. Gemini provided a 'bridge to the Moon' by developing the procedures for space rendezvous and docking vital for the Apollo LOR missions.

◀

Apollo 11 astronauts Buzz Aldrin and Neil Armstrong practise their moonwalk activities in April 1969. Simulating the Moon's one-sixth gravity proved to be especially difficult.

The first creature to go into space was a Russian dog called Laika. Only one month after the launch of Sputnik 1, Laika launched aboard Sputnik 2, travelling in a padded, pressurised compartment. With no hope of a return to Earth, she survived for only a few hours according to some sources. The Americans preferred to send monkeys, most famously Ham, who flew in the MR-2 Mercury mission in January 1961 and was recovered in excellent condition after splashing down in the Atlantic.

➤

The Lunar Landing Training Vehicle, or 'flying bedstead' as it became known, was intended to mimic the flight characteristics of a lunar lander. It was an unforgiving machine and on one occasion Neil Armstrong ejected with only seconds to spare.

the lower descent stage, and then return them via the upper ascent stage.

To loft the Apollo hardware into space a team led by the former German rocketry expert Wernher von Braun had developed the mighty Saturn V, a 363ft (110.6m)-tall rocket consisting of three stages: the first stage, the S-IC, had five F-1 engines which burned liquid propellants to generate 7.5 million pounds of thrust and accelerate the

Did you know?
The name for the Apollo programme was suggested by Abe Silverstein, an early director of the Lewis Research Centre. The ancient Greeks regarded Apollo as a god who accompanied travellers on their way.

Charlie Duke and John Young rehearse with a Lunar Rover test vehicle on a simulated lunar surface in preparation for Apollo 16.

15

spacecraft to around 6,000mph (2.68km/s). Next came the second stage, S-II, with five engines taking it to 15,300mph (6.84km/s) and an altitude of 115 miles (185km). Once depleted of fuel, the third stage, the S-IVB, took over to carry the Apollo payload into orbit; this final stage was designed to be restarted for the outward push to the Moon.

Even as Apollo was being shaped, the one-man Mercury and two-man Gemini missions were establishing the stepping-stones to the Moon, ironing out the techniques for prolonged spaceflight and honing the procedures for the multiple rendezvous and docking manoeuvres necessitated by the LOR mission plan. By the time of Gemini's conclusion in November 1966, the first manned flight of the new Apollo spacecraft was only three months away.

Apollo 1 was to be a shakedown mission to thoroughly test all aspects of the prototype command and service modules in Earth's orbit. NASA's approach to the Moon landings was to make a series of incremental steps, carefully evaluating equipment and procedures until everything was ready for the real thing. That was the theory, at least, but keeping to Kennedy's end-of-decade deadline was putting on the pressure.

Added to this was the question of whether the Soviets were ahead of them. It seemed that every time the Americans took one step forward, the Soviets took two. If the USA proposed sending two men into orbit with Gemini, the Soviets sent three in Voskhod 1. If they planned a rendezvous or a space walk, the Soviets did it first. In fact, the Soviets were playing a clever game, responding to NASA's openness in announcing its intentions and then stealing their thunder. The Voskhod capsule, for example, was intended for a two-man crew, but they squeezed a third man in by getting rid of the bulky spacesuits. In truth both sides were in danger of pushing too fast, with safety being compromised in the name of political expedient.

On Friday 27 January 1967 the crew of Apollo 1 strode across the gantry at Launch Pad 34 to the White Room which

Gemini veterans Ed White and Gus Grissom pose with rookie astronaut Roger Chaffee and a model of the ill-fated Apollo 1 capsule.

The crew of Apollo 1 cross the launch pad gantry to the White Room on the day of the fire, 27 January 1967.

inexplicably blew after splashdown and the capsule sank. Despite Grissom's strenuous denials that he had triggered the hatch bolts, here was a man with something to prove. Accompanying Grissom were Ed White, the first American to walk in space, and rookie astronaut Roger Chaffee.

As Grissom entered the White Room he gave the Apollo capsule an involuntary look of disdain. His concerns were prompted by a string of hitches and last-minute modifications, and at one point he openly vented his feelings by hanging a lemon on the flight simulator. Now they were clambering aboard the real thing for a full dress rehearsal of pre-launch procedures. This involved disconnecting external power supplies, pressurising the cabin with a 100 per cent oxygen atmosphere and bolting the hatch shut – in effect placing the

surrounded the capsule perched on top of a 200ft-high Saturn 1B booster. Heading the trio was Commander Gus Grissom, a veteran of Mercury and Gemini. He had been the second American in space aboard Mercury 4's brief sub-orbital lob; a mission which ended under a cloud when a hatch

'As test pilots must be, we are accustomed to death. The loss of a colleague, often a good friend, is not an uncommon occurrence. It doesn't mean we don't feel deeply about our friends and don't mourn their passing... But we don't wear a black arm band forever. We mourn the man for a little while, then we live with his loss.'

Wally Schirra, writing about the Apollo 1 fire

Did you know?
The entire surface of the Moon is pock-marked with impact craters, with approximately 1 million having a diameter greater than 0.54 miles (1km). The largest is on the far side and measures a massive 1,210 miles (2,240km) in diameter.

capsule in self-sufficient mode. Grissom had also requested a rehearsal of capsule 'egress', or escape procedures, for later in the day.

Things went badly from the start; Grissom detected an unpleasant odour from the cabin's life-support system. Then there were problems with the communications link, causing him to snap at the Control Room; 'If I can't talk to you five miles away how can we talk to you when we are on the Moon?'

The interior of the Apollo 1 capsule was so badly damaged by the intensity of the fire that rescuers couldn't even find the three bodies at first.

After more than five hours of this the astronauts were ready to proceed with the final stage of the countdown from T-minus ten minutes.

Suddenly a voice was heard over the intercom, 'Fire! We've got a fire in the cockpit!' It had started by Grissom's feet at the bottom of the left-hand couch, most likely sparked by a faulty wire. The fire rapidly took hold in the oxygen-rich environment. Six seconds into the incident instruments recorded an abrupt rise in both pressure and temperature. Saturated with oxygen, the cushioning covering the walls, even the netting designed to stop loose items floating about, all burned with the ferocity of a blowtorch. Grissom and White struggled to get the two-piece hatch open, but opening-inwards it was held tight by the mounting internal pressure

Did you know?

Just a few months after the Apollo 1 fire, the first flight of the Soviet's new Soyuz spacecraft ended in disaster. Cosmonaut Vladimir Komarov was killed when Soyuz 1 suffered a series of malfunctions and plummeted back to Earth beneath a tangled parachute. Before leaving orbit, and knowing his probable fate, Komarov was allowed a heart-rending final conversation with his wife.

and six securing sockets. Even in the best of conditions it might have taken them a minute and a half to get it open. In these conditions it was impossible.

Having failed to open the hatch, Grissom and White tried to shield themselves underneath the couches, while Chaffee, furthest away from the fire, stayed in his couch to maintain communications. 'We've got a bad fire. Let's get out. We're burning up...' The build-up in pressure split the capsule's hull, while in the White Room the technicians saw movements through the small hatch window; a white helmet followed by the flash of flame. It was over so quickly, with just fourteen seconds between the first cry of alarm and the last. The crew of Apollo 1 had slipped into unconsciousness from the noxious fumes, and death soon followed.

As experienced pilots, all astronauts are aware of the dangers they face in the air, but none expect to die in a fire on the ground. Grissom, White and Chaffee had paid the ultimate price. The only consolation was that the Apollo programme would be a safer and more successful one because of their sacrifice.

Nearly twenty months elapsed before another astronaut flew in Apollo. In the interim there were several unmanned test launches (Apollos 2-6), so by the time NASA got to their next manned mission the numbering was way up to Apollo 7, which blasted-off on 11 October 1968. Apollo 7's mission was to put the redesigned CSM through its paces, to rebuild confidence in the programme and get America back on course for the Moon.

In command was one of NASA's most experienced astronauts, Wally Schirra, the only man to fly on Mercury, Gemini and Apollo missions. With him were newcomers Walter Cunningham and Don Eisele. From the start Schirra's team took a close interest in every aspect of the new capsule's design and construction. There had been several significant modifications following the recommendation of the Review Board investigating the Apollo 1 fire. Most noticeably the hatch was redesigned to

The launch of the Apollo 7 two-stage Saturn 1B booster from Cape Kennedy on 11 October 1968. The 1B is smaller than the Saturn V, its second stage serving as Saturn V's third stage.

➤

Prime crew for Apollo 7: Don Eisele, Walter or 'Wally' Schirra and Walt Cunningham. Schirra joked that they were known as 'Wally, Walt and what's-his-name'.

'If we were to start today on an organised and well-supported space program, I believe a practical passenger rocket can be built and tested within ten years.'

Dr Wernher von Braun, 1955

open outwards, and, with crew safety now paramount, all flammable materials were replaced. Invisible to the eye, but just as crucial, the practice of pressurising the cabin with 100 per cent oxygen for ground tests was abolished in favour of a 40/60 mix of nitrogen and oxygen at normal atmospheric pressure.

With no LM available for Apollo 7, they would fly a repeat of Apollo 1's intended mission. Many considered Schirra to be a pair of safe hands for this important return to space, although his forthrightness was sometimes in conflict with the flight managers. Before lift-off he challenged them over a decision to launch in questionable wind conditions, as he felt that the capsule was not suited to a landing on the ground if there was an abort. In the event the Saturn 1B rocket performed impeccably, placing the spacecraft in an elliptical orbit ranging from 173 to 140 miles. 'We're having a ball... she's riding like a dream,' confirmed Schirra in a slightly happier mood.

The CSM remained attached to the second stage of the Saturn 1B until the end of the second orbit. Apollo 7 would practice station keeping with the empty stage as a target for simulated dockings.

For the first time in the American space programme the astronauts were able to remove their bulky pressure suits and enjoy

On a lunar mission the shroud panels at the top of the third stage protected the folded lunar module (LM) housed inside. Apollo 7 had no LM and because one of the panels had not opened fully they were to be jettisoned on future missions.

the greater comfort of the Apollo capsule. There were hammock spaces underneath the left and right couches, and they even had hot water for the preparation of food, but none of this was enough to raise Schirra's mood. His determination to turn in a picture perfect mission kept his crew on a strict itinerary and he wasn't going to let anyone, not even Mission Control, get in the way of that. Frictions surfaced twenty-six hours after launch when he refused to switch on the cameras for a live-from-space broadcast. 'The show is off. The television is delayed without further discussion. We've not eaten, I've got a cold and I refuse to foul up our time in this way.' In fairness a head cold in zero-gravity is the last thing you need as the mucus can't drain from the sinus passages.

Otherwise the mission was going well. The main engine at the rear of the CSM, the Service Propulsion System (SPS), performed perfectly. By the third day the 'Wally, Walt and Don' TV special went ahead with the crew floating about and holding cards up to the camera, 'Hello from the lovely Apollo room high above everything' followed by,

Did you know?

Star sailors: the term 'astronaut' is derived from two Greek words, *astro* for star and *nautes* for sailor, while the Soviet/Russian equivalent of 'cosmonaut' features the Greek *kosmos* meaning universe. Chinese space travellers are widely referred to as 'taikonauts', from the Chinese word *taikong*, meaning space. Naturally the French favour their own 'spationaute', which is derived from the Latin *spatium* for space.

'Keep those cards and letters coming in folks'.

But elation descended into a series of petty squabbles with Mission Control, a mix-up over sleep times, complaints about scheduling further broadcasts and so on. Trapped in their tin can, the three men were feeling lousy because of their colds, the food was poor and halfway through the mission they were getting bored of just orbiting the Earth, partly because Schirra had banished all books, magazines and tapes as part of the regime to rid the capsule of flammable substances. Matters came to a head with the approaching re-entry. Schirra, concerned that their medical conditions could result in permanent damage to their eardrums, wanted to do away with their helmets on re-entry so they could hold their noses and blow to relieve the pressure. Mission Control was not happy with this, but eventually gave way.

Thankfully, the command module and its heat shield functioned perfectly for the splashdown south-east of Bermuda, only a mile from the designated point. While Apollo 7 had not been a happy ship and none of the crew flew in space again, the verdict on the redesigned spacecraft was that it was '101 per cent successful'.

Apollo 7 commander Wally Schirra looking out of the rendezvous window on the ninth day in orbit. He was nursing a head cold, a particularly unpleasant condition in zero gravity as the sinuses cannot drain normally.

A spectacular view of the Florida peninsula seen from orbit aboard Apollo 7.

The objectives of each successive mission, and indeed the crew to fly them, were in a state of constant review. Apollo 8, was originally slated as an Earth orbit test of the Saturn V booster and Apollo spacecraft complete with lunar module, but given the flawless operation of Apollo 7 and delays in delivering the first LM, NASA swapped missions with Apollo 9, gambling on an ambitious flight around the Moon.

With the comfort of twenty-twenty hindsight it is all too easy to underestimate the massive leap of faith this decision represented. This was a full-on mission to the Moon, albeit without an actual landing, but bringing with it a host of new challenges. Firstly, it was the first manned flight with the Saturn V booster, and only the third time it had been test flown at all, with some significant issues still to

> ➤
> *James Lovell, Bill Anders and Frank Borman of Apollo 8 – the first mission to encircle the Moon.*

be resolved. Secondly, it entailed several critical engine burns, initially to send the spacecraft to the Moon, then to establish it in orbit and finally to kick it back on the correct trajectory for Earth. On each occasion they had to function perfectly, and, as if that wasn't enough, the latter two burns would be instigated by the crew while behind the Moon and out of contact with Mission Control.

This swapping of missions also entailed a reshuffling of crew, and consequently

The five F1 engines of the Saturn V S-1C first stage, during final assembly in the Vehicle Assembly Building (VAB). Apollo 8 was the first manned launch of the Saturn V.

it was commanded by Frank Borman who was accompanied by fellow Gemini veteran Jim Lovell and first-timer Bill Anders.

Did you know?
Each stage of the Saturn V launch vehicle was constructed by a different company. Boeing built the big S-IC first stage, North American Aviation the S-II second stage, and Douglas Aircraft the S-IVB third stage.

Bang on schedule, Apollo 8 lifted-off on 21 December 1968 to attain Earth orbit. At a little over two and a half hours into the mission they got the signal, 'Apollo 8, you are Go for TLI'. The TLI, or Trans-Lunar Injection, entailed firing the third-stage engine for over five minutes and accelerating the spacecraft to 24,200mph (38,950km/h), the fastest any human had ever travelled.

From there on Apollo 8 would coast to the Moon, but not before completing another manoeuvre. Although there was no LM to extract from the housing at the top of the redundant third stage, the CSM was detached and slowly rotated 180 degrees to point backwards for a simulated docking.

As the Earth's gravitational pull began to slow Apollo 8 down, the crew settled in for a long cruise. To ensure that the heat from the Sun was evenly distributed on the hull, they initiated a slow 'barbecue' roll. Now was the time to catch up on some sleep at last, although Borman was worrying the medical experts with a display of flu-like symptoms including nausea and vomiting. Approaching the halfway mark, the now healthy crew transmitted the first pictures of the bright Earth set against impenetrable

blackness. Tantalisingly blurry at first, the later broadcasts showed near perfect images which were seen by millions. Lovell described the scene: 'The Earth is now passing through my window. It is about as big as the end of my thumb. Waters are all sort of a royal blue; clouds, of course, are bright white...'

As the Moon drew ever nearer, the astronauts prepared for the first burn of the CSM's main engine to begin Lunar Orbit Insertion (LOI). In essence this was a breaking manoeuvre carried out behind the Moon to slow them down and maintain lunar orbit. Failure at this stage could be catastrophic as the spacecraft would either drift off into space or crash into the lunar surface. At Mission Control it was a particularly anxious wait for radio contact to resume. After what seemed like an

'You could see the flames and the outer skin of the spacecraft glowing; and burning, baseball-size chunks flying off behind us. It was an eerie feeling, like being a gnat inside a blowtorch flame.'

Bill Anders on the Apollo 8 re-entry

eternity the speakers finally crackled into life to affirm success.

Once established in lunar orbit, the astronauts had an unprecedented view of the surface below them. 'The Moon is essentially grey, no colour,' said Lovell. 'Looks like plaster of Paris. Sort of greyish sand...' Then, on Christmas Eve, they sent what would prove to be one of the most evocative messages ever broadcast from space. It was Anders who introduced a reading from the book of Genesis; 'In the beginning, God created the heaven and

◄
Christmas Eve 1968: as they emerged from behind the Moon the Apollo 8 astronauts were greeted by the first sight of an Earthrise. Strictly speaking this should be orientated ninety degrees to the left, but the human eye naturally seeks a conventional horizon.

The Apollo 8 astronauts arrive on board the recovery vessel USS Yorktown, *following a splashdown in the central Pacific Ocean on 27 December 1968.*

the Earth...' Each astronaut took a turn, and then Borman concluded the transmission. 'From the crew of Apollo 8 we close with goodnight, good luck, a Merry Christmas, and God bless all of you – all of you on the good Earth.'

There was one final big burn to come. After ten orbits the main engine was successfully fired to perform the Trans-Earth Injection (TEI). In other words they were going home.

By early 1969 there was a feeling of optimism in NASA; a sense that Kennedy's end-of-decade deadline was achievable. This was in no small part due to the smooth flights of Apollos 7 and 8; if the next two missions adhered to expectations, it was possible that Apollo 11 might make the first landing, or, failing that, Apollo 12.

Either way, time was short and both Apollo 9 and 10 were going to be enormously demanding missions.

On 3 March 1969 the Apollo 9 crew blasted-off from Cape Kennedy to tackle Apollo 8's original mission of testing the LM in low Earth orbit. Riding onboard the most heavily laden Saturn V so far were James McDivitt, David Scott and Russell 'Rusty' Schweickart, while stored within the top of the third stage was the long-awaited lunar module. An extraordinary bundle of angular components, smothered with bristling antennae and perched on four spindly legs, the LM had no regard for the niceties of aerodynamics as it was designed to operate only in the vacuum of space. It consisted of two main sections; the lower descent stage with its main engine to land on the Moon, and the upper ascent stage

Apollo 9 line-up, James McDivitt, Dave Scott and Russell 'Rusty' Schweickart. This mission was designed to test the Apollo CSM and LM in Earth orbit.

with another engine to thrust the astronauts back into lunar orbit and the awaiting CSM. To save on weight its skin was little more than tin foil and a wayward screwdriver could have easily punched a hole right through it. When Jim McDivitt first clapped eyes on the LM at Grumman's Works he described it as a 'tissue paper spacecraft'.

➤ Shown in Grumman's clean room, the Apollo 9 lunar module, or LM – usually pronounced as 'Lem' after the earlier designation Lunar Excursion Module.

Shortly after Apollo 9 attained orbit, David Scott fired the explosive bolts to separate the CSM from the rocket's third stage, and to remove the panels surrounding the LM. Then he nursed the CSM through 180 degrees to face a docking drogue on the top of the LM and, ever so gently, the two craft were locked together. Once the docking tunnel had been pressurised the forward hatch could be opened, electrical power linked up and the two spacecraft became one. Another box ticked.

On day three, despite a bout of space-sickness, Schwieckart donned his spacesuit to spend thirty-eight minutes on the porch

The lunar module Spider, *ready for extraction from the top of the Saturn V IVB third stage. The protective panels have already been jettisoned.*

of the LM while Scott filmed him from the open hatch of the CM. It had been proposed to demonstrate the possibility of an astronaut making his way from the LM to the CM in the event of an emergency, but this aspect of the EVA was scrubbed because of Schweickart's earlier space sickness. Even so, it was a successful spacewalk and the first using a self-

contained life-support system rather than an umbilical connection to the spacecraft. Another tick.

With two spacecraft in operation, NASA allowed the astronauts to devise their

Dave Scott stands at the open hatchway of the Apollo 9 capsule, photographed by Rusty Schweickart standing on the porch of the lunar module.

◄
The Spider *has wings: The 'golden slippers' foot restraints, which secured Schweickart during his EVA, are visible on the LM's porch.*

◄◄
Rusty Sckweickart seen through the LM's window. This was the first test in space of the Apollo A7L spacesuit with its independent life-support backpack intended for the lunar EVAs.

Did you know?

Unlike its successor, the Space Shuttle, none of the Apollo hardware was designed to be reusable. The first and second stages of the Saturn V fell back into the Atlantic Ocean, while the third stage was usually fired into orbit around the Sun. Only the tip of the stack, the CM capsule, could re-enter the Earth's atmosphere.

against the blue arc of the good Earth as a series of tentative test firings of the altitude thrusters tipped it this way and that. Then it was time to fire-up the main engine of the descent stage to watch the fireworks. A splatter of yellow flame erupted from the bell-shaped nozzle, shooting *Spider* to a distance of 113 miles (182km) from the security of *Gumdrop* – their only ride back to Earth.

The final test for the LM was to separate the ascent stage, in which the astronauts were riding, from the descent stage which, on an actual Moon landing, would serve as its launch platform. All went well and as Schweickart and McDivitt manoeuvred their craft back within sight, Scott greeted his companions; 'You're the biggest, friendliest, funniest-looking spider I've ever seen.'

> The ascent stage Spider returns to Gumdrop on the fifth day in orbit. It is pointing downwards in this photograph and displays the single ascent engine.

own call-signs to avoid confusion. The flying bug became *Spider*, for obvious reasons, while the CM was *Gumdrop*. On day five it was time to separate *Spider* and *Gumdrop*, with McDivitt and Schweickart onboard the LM. For a while it danced

Less than two months later and another lunar module was flying through space, this time within a hair's breadth of the Moon.

Apollo 10 had no room for rookies, just three experienced space travellers; Tom Stafford in command, John Young as CM pilot and Gene Cernan as the LM pilot. This was a dress rehearsal for the first landing, flying within 47,000ft (15.6km) of the cratered surface of the Moon to test navigational equipment and reconnoitre potential landing sites. The launch from Cape Kennedy on the afternoon of 18 May 1969 was mostly routine stuff, although the ride into orbit had been an exceptionally bumpy one. The journey between Earth and the Moon was uneventful, and it wasn't until lunar orbit that trouble really started.

On day four Stafford and Cernan were in the LM 'Snoopy' and drifting away from Young in the CM, good old Charlie Brown. The separation had taken place on the far side of the Moon, out of contact with Earth, and in that moment Young became the most isolated man alive, knowing only too well that if anything went wrong with the LM he would have a very long and lonely return to Earth without his colleagues.

◄

Gene Cernan and Tom Stafford training in the LM simulator.

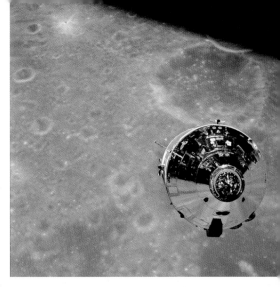

➤
The Apollo 10 command module pulls away from the lunar module Charlie Brown *shortly after separation.*

Did you know?

The Moon is our closest neighbour by a long shot. The mean distance between the Earth and the Moon is around 240,000 miles (386,200km) while Mars is 35 million miles (56,000,000km) away at its nearest. That's 150 times further than the Moon – a very long way to send a manned mission.

As they came back around the rim of the Moon the two craft were 50ft (15m) apart, with Young capturing the LM's progress on a colour television camera, another Apollo first. Final checklists completed, it was time to fire up *Snoopy*'s descent stage engine, and the LM shrank away into the distance to trace an invisible line above the lunar surface.

Snoopy swooped low before levelling out; the craters and boulders expanded into stark relief as they made the first pass across the most probable landing spot for Apollo 11, Mare Tranquillitatis – the Sea of Tranquillity. 'I tell you we are low, we are close, babe... Down there! The landing site!'

On a final loop around the Moon they climbed to the highest point of their orbit and made a steep run back over the Sea of Tranquillity in preparation for a simulated blast-off from the surface. *Snoopy* divided in half and the ascent stage engine fired to hurtle them back into orbit for a rendezvous with the CM. Part of the procedure was to test the backup navigational system, but as luck would have it the astronauts

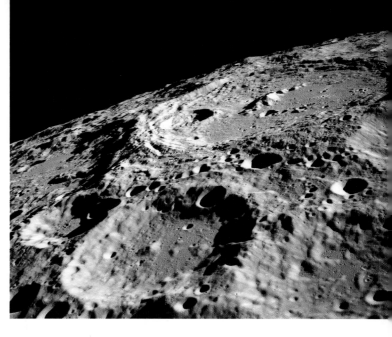

had flicked a vital switch into the wrong position. When the bolts were fired to separate the two stages all hell broke loose. Suddenly diving and spinning totally out of control, Stafford instinctively flipped the switch, thinking he was re-engaging the main navigation computer but actually compounding the problem further. The radar which should have been seeking the CM was sending them on a course straight for the Moon.

Stafford left Mission Control in no doubt about their situation. 'We're in trouble...' he radioed. '*Snoop*, we show you close to gimbals lock!' came the response. If

A large terraced crater on the lunar limb, photographed from Apollo 10.

45

The first international handshake in space took place during the last ever Apollo flight, the Apollo-Soyuz Test Project of 1975, when Americans and Soviet astronauts linked up in Earth orbit. This marked the start of a new era of cooperation between the two nations and paved the way for the International Space Station.

the gyroscopes orientating the navigational computers in relation to the lunar horizon overloaded then it was all over. Fearing that a thruster had stuck open, Stafford took manual control and swiftly had the spacecraft back on an even keel. Incredibly, this whole episode only lasted fifteen seconds; a lifetime to the astronauts who had escaped with only moments to spare.

Despite these distractions, the ascent stage engine fired bang on schedule and after three hours apart they were reunited with the CM. It was business as usual for the commander of Apollo 10 as he calmly radioed Mission Control, '*Snoopy* and *Charlie Brown* are hugging each other.'

Looking back on the flight and whether he had been disappointed not to make the first landing with Apollo 10, to have been so near and yet so far, Gene Cernan wrote,

Charlie Brown *prepares to dock with* Snoopy.

A *helicopter from USS* Princetown *picks up the crew of Apollo 10.*

'We all believed in the importance of our mission because we knew Apollo 11 was going to need every scrap of information we could gather...'

The *Eagle* was at 2,000ft (610m) and on its way to making the first lunar landing ever when the alarm sounded. 'Programme alarm, twelve-oh-one,' Buzz Aldrin called out to mission commander Neil Armstrong, informing him it was a navigation computer overload. Armstrong concentrated on piloting their improbable flying machine, looking ahead to the intended landing site. At a height of 1,000ft (305m) he could make out a field of jagged boulders directly in the way and he pitched the lander forward toward clearer ground beyond. At 300ft (92m) the fuel was down to 8 per cent as they chased their shadow across the barren surface. Now there was a crater in the way. He nudged the hand controller forward, levelled out and spotted a clearer area beyond that. With ninety seconds of fuel left the dust began to kick up around them. Sixty seconds. The engine blast was throwing up so much dust he could barely get a fix on the ground. Thirty seconds of fuel left... It was a close call.

➤

The classic image of Man on the Moon: Neil Armstrong and the lunar module are reflected in Buzz Aldrin's visor during their EVA on the Sea of Tranquility.

At Mission Control the tension was tangible as they waited for word from Apollo 11. 'Houston, Tranquillity Base here. The *Eagle* has landed.' Less than eight short years since America had sent a man into orbit they had accomplished the greatest technical challenge mankind had ever faced. On 20 July 1969 they had reached another world.

Neil Armstrong suiting-up for the Apollo 11 mission.

Unbelievably the mission planners had scheduled a four-hour sleep period before the first Extra Vehicular Activity (EVA) on the Moon's surface, but in the event Armstrong and Aldrin called for the

The start of the journey: Neil Armstrong, Michael Collins and Buzz Aldrin head for the crew bus on their way to the launch pad.

moonwalk to be brought forward. As he descended the ladder, Armstrong reached across to release a small camera to transmit live pictures to Earth. At the bottom rung he paused, clearly visible to an estimated billion viewers, and took the last 3ft (1m) gap in a slow-motion-bound landing with his feet still on the LM's footpad. 'The surface appears to be very, very fine grained as you get close to it; it's almost like a powder...

The moment of blast-off on 16 July 1969, sending Apollo 11 on its way to the Sea of Tranquility.

Okay, I'm going to step off the LM now.' And with one foot on the Moon and the other on the footpad he uttered the historic words, 'That's one small step for [a] man, one giant leap for mankind.'

Weighing a mere 60lb (27kg) in the gentle one-sixth gravity, Armstrong made his first tentative steps, learning how to move in the new conditions despite the restrictions of his spacesuit. The immediate task was to collect a small sample of soil and stash it away in a pouch, a contingency sample in case the moonwalk was aborted. Then came Aldrin's turn to accompany him on the surface, and after a momentary loss of words he described his new world as 'magnificent desolation'.

The astronauts moved the camera further away from *Eagle*, transmitting ghostly images as they moved about,

Did you know?

In 2002 a new asteroid appeared in Earth orbit and was designated as J002E3. It was subsequently identified as the Apollo 12 S-IVB third stage and has since returned to a solar orbit.

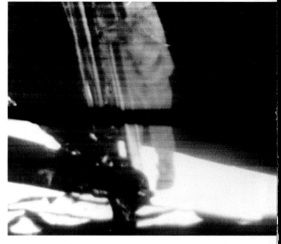

Tranquility Base: millions of viewers watched the live television images of Neil Armstrong climbing down the ladder to the surface.

Composite image showing Buzz Aldrin descending to the foot of Eagle's steps.

their uncertain steps evolving into a more confident bouncing lope. Armstrong read aloud from a plaque attached to the descent stage. 'Here men from the planet Earth first set foot on the Moon, July 1969, A.D. We came in peace for all mankind.' Then they planted the stars and stripes – held out in the airless vacuum by a metal spring – putting to rest JFK's nightmare vision of a red flag on the Moon. After two and a

'I just feel lucky to have been born in 1930. I grew up with biplanes and *Buck Rogers*, learned to fly in the early jet fighters, and hit my peak when Moon rockets came along. Before I die, who knows what I may see?'

Michael Collins, CM pilot Apollo 11

Aldrin salutes the flag, held aloft in the air-less vacuum by a metal spring wire.

Moon was the maiden name of Buzz Aldrin's mother. The name Buzz came from his sister's childhood mispronunciation of the word 'brother'. He made it his legal name in 1988.

➤

The small flag-like device is the Solar Wind Composition (SWC) experiment to register impacts from millions of tiny solar particles.

half hours it was time to re-enter *Eagle*. They had collected around 48.5lb (22kg) of lunar soil and rock samples, conducted a number of experiments including setting-up instruments to register seismic activity and the flow of radiation particles from the Sun. They even found time for a brief chat with President Nixon via a direct link to the White House.

Twenty-one hours after their arrival on the Moon, the ascent stage rocketed them away from the lunar surface for a

After the EVA a weary Neil Armstrong grins with satisfaction.

Buzz Aldrin photographed in the LM before the landing.

Back in lunar orbit, the ascent stage of Eagle *closes in on Michael Collins in the command module,* America.

Once aboard the recovery ship, USS Hornet, *the returning astronauts were held in quarantine for three weeks. President Nixon was on the ship to welcome them back in person.*

rendezvous with Michael Collins waiting patiently in the CM *Columbia.* Sixty hours later they were back on Earth; a planet forever changed by what they had done. Buzz Aldrin would later sum up the significance of Apollo 11's pioneering mission: 'This has been far more than three men on a voyage to the Moon... This stands as a symbol of the insatiable curiosity of all mankind to explore the unknown.'

As Pete Conrad stepped off the lower rung of the LM's ladder onto the Moon's Ocean of Storms, he exclaimed, 'Whoopie! Man, that may have been a small one for Neil, but that's a long one for me.'

If it hadn't been for the reshuffling of Apollo 8 and 9 bumping his crew back one mission, Conrad might have commanded Apollo 11. And if either Apollo 9 or 10 hadn't gone so incredibly smoothly, Apollo 12 could have been the first to make the lunar landing. But it was Conrad's fate to command the follow-up mission, and history has a cruel way of passing over those who come second. As the public and media saw it, America had already landed a man on the Moon, so what was the big deal about going back there? To make matters worse a malfunction of the television equipment made this the one landing without live pictures. Apollo 12 wasn't only in the shadow of its illustrious predecessor, it went almost unnoticed.

Pete Conrad, Gordon Cooper and Al Bean – the crew of Apollo 12 for the second lunar landing mission in November 1969.

Did you know?

The first three batches of astronauts returning from the Moon were placed in quarantine for two to three weeks in case of contamination from unknown bacteria. The Apollo 14 crew was the last to be quarantined.

The most dramatic moment of the mission occurred during the launch, on 14 November 1969, which went ahead despite heavy rain storms in the Cape Kennedy area. On board with Conrad were Richard Gordon, CM pilot, and Alan Bean, LM pilot. After 36.5 seconds the 363ft (111m)-high Saturn V became a massive lightning conductor when a bolt struck, plunging the capsule into momentary

'In spite of the opinions of certain narrow-minded people who would shut up the human race upon this globe, we shall one day travel to the Moon, the planets, and the stars with the same facility, rapidity and certainty as we now make the ocean voyage from Liverpool to New York.'

Jules Verne – *From the Earth to the Moon*, 1864

darkness and causing a break of several seconds in the telemetry feeds to Mission Control. The backup batteries came on

◀
Conrad and Bean during a pre-flight run-through of their lunar surface activity.

Did you know?

In 1970 the Soviet Union landed the unmanned Luna 16 probe on the Moon where it scooped up 22lb (10kg) of samples before returning to Earth. This led many sceptics to question why the USA needed to send men.

line and Pete Conrad voiced his concerns. 'I think we got hit by lightning. We just lost the guidance platform, gang. I don't know what happened here.' Thankfully the Saturn V's independent guidance system had not been affected by the jolt, and with telemetry restored an abort was narrowly avoided. Riding a column of flame, Apollo 12 pushed on up into a parking orbit where the crew were able to verify the functionality of their spacecraft before initiating the Trans-Lunar Injection (TLI).

Unlike Apollo 11's over-run, the landing of Apollo 12's *Intrepid* was an exercise in precision, touchdown being within 600ft (185m) of the unmanned Surveyor 3 probe which had landed thirty-one months earlier. Any closer and there were concerns that the descent engine would throw dust over Surveyor.

As Conrad and Bean began their first lunar walk on 19 November, Gordon encircled the Moon in *Yankee Clipper*. Because of the accuracy of the landing he was able to pinpoint their position and excitedly informed Mission Control, 'I have *Intrepid*'. Then, spotting a second point of light: 'I see Surveyor! I see Surveyor. That's almost as good as being there.' Gordon had a better view than the television audience back home because Bean had inadvertently pointed the camera directly at the Sun, wrecking the vidicom tube. From that point onwards this would be a sound-only show. Undaunted, the two astronauts continued their work in a light-hearted mood, with Pete Conrad humming away to himself. At one point he commented, 'I feel like I could stay out here all day.'

Conrad and Bean made two highly successful moonwalks during their two-day stay, with a total duration of almost seven hours – about three times that of their predecessors – collecting 75lb (34kg) of rocks and setting up instruments to measure the Moon's seismic activity, solar wind flux and magnetic field. The surface conditions were noticeably different

▲
Al Bean exits from Intrepid *to join Conrad on the surface.*

Bean holds a special sample container at Sharp Crater. Notice the EVA check-list clearly visible on his left wrist.

and their spacesuits like coal dust, to the extent that they decided to remain in them during sleep periods back in the LM, so the dust wouldn't clog up the suits' connectors.

The highlight of the mission came on the second day with a walk over to the shallow crater where Surveyor 3 was waiting. The astronauts removed samples of the tubing and the camera, to be taken back for analysis to show how the materials had survived their time on the Moon. While there they took time to pose like tourists, although, sadly, Bean could not find the self-timer device he had smuggled aboard with the intention of photographing himself together with Conrad by Surveyor. He had wanted to play a prank on the post-mission photo analysts as they struggled to work out how the picture could have been taken.

from those of Apollo 11. 'Dust got into everything,' Conrad explained. 'You walked and a pair of little dust clouds kicked up around your feet.' It clung to their boots

In a more sombre moment Bean laid Clifton Williams' pilot wings on the lunar soil as a tribute to a fellow astronaut who had died in an aeroplane accident. Williams had trained with Conrad and Gordon as backup on Apollo 9, and would have been assigned as LM pilot on Apollo 12. It was a poignant reminder that this was a perilous business, as the crew of Apollo 13 were about to find out.

Unlucky thirteen. By the time Jim Lovell, Fred Haise and Jack Swigert lifted-off at 13.13 CST on 11 April 1970, the world had tired of the Apollo flights, and their mission to the Fra Mauro uplands was greeted with indifference. Such was the lack of interest that when they made a live broadcast midway through the third day of their journey, few networks ran with it. 'This is the crew of Apollo 13 wishing everyone

Jim Lovell, Tom Mattingly and Fred Haise were the original crew for Apollo 13, but Mattingly was replaced at the last minute by Jack Swigert, following his exposure to German measles.

there a nice evening,' concluded Lovell in the congenial and confident tones of an old space hand. 'We're just about ready to close out our inspection of *Aquarius* and get back for a pleasant evening in *Odyssey*. Goodnight.' Shortly afterwards the mission was turned on its head and in the ensuing days the plight of Apollo 13 would captivate the entire world.

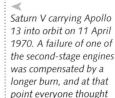

Saturn V carrying Apollo 13 into orbit on 11 April 1970. A failure of one of the second-stage engines was compensated by a longer burn, and at that point everyone thought the glitches were behind them.

Shortly before disaster struck, Mission Controllers watch Fred Haise via the live TV broadcast from Apollo 13 on its way to the Moon.

The view from Aquarius as it passed around Lovell's 'lost Moon'.

Haise was in the LM, *Aquarius*, with Lovell tidying up the camera and wires midway, while Swigert went about some routine business in the CM, *Odyssey*. As the backup CM pilot he had only joined the crew a couple of days before lift-off because Thomas Mattingly had been exposed to German measles. When an amber light indicated low pressure in a tank located in the cylindrical service module (SM), Mission Control called him up. 'We'd like you to stir up your cryo tanks.' The liquid hydrogen and oxygen tanks were fitted with fans to prevent stratification in zero gravity conditions, but moments after Swigert threw the switches the spacecraft was shaken by a bang. 'OK, Houston, we've had a problem.'

Pitching up and down, out of control, Lovell reported that the oxygen pressure for Tank 2 was reading zero and the pressure in Tank 1 was falling before his eyes. Had they been hit by a meteorite?

Without power and oxygen in the CM, the only option was to use *Aquarius* as a lifeboat. Mission Control instructed the crew to shut down *Odyssey*, to conserve its meagre battery reserves for the vital re-entry manoeuvre, and to move into *Aquarius'* cramped cabin. Unfortunately this was designed to maintain two people on the Moon for up to fifty hours, not to sustain three men for almost four days – the time it would take to get them back to Earth. A turnaround and direct flight home was impossible with the extent of the damage to the SM unknown, so instead the main engine of the *Aquarius'* ascent stage was fired up to put Apollo 13 on the correct course for a slingshot around the Moon and a free-trajectory home.

For the two rookie astronauts their swoop around the Moon was a time of

Did you know?
The first man-made object to reach the Moon was the Soviet-built Luna 2 probe, which impacted on the surface in 1959. Seven years later the Luna 9 successfully touched-down on the Ocean of Storms, proving that a lander would not sink into the lunar dust.

◄

One of only a few photographs taken during the return flight, it shows Haise and Swigert with the jury-rigged pipework for the square 'mail box' which contained the lithium hydroxide CO2 scrubbers.

After separation from the CM and LM, the astronauts had only a few moments to grab photographs to show the extent of the damage to the stricken Service Module. An entire panel has been blown away by the explosion.

Following a splashdown in the South Pacific, Haise, Lovell and Swigert stepped from the helicopter onto the decks of the USS Iwo Jima.

wonder as they gathered at the windows to photograph this desolate alien world. For Lovell it was a bitter-sweet moment, a second visit to what he later referred to as his 'lost Moon'.

While conditions inside *Aquarius* were uncomfortable – with temperatures barely above freezing, condensation soaking their clothes and drinking water restricted to only six ounces per person, per day – they were survivable. But without clean air to breathe the astronauts were in danger of suffocating in their own carbon dioxide. *Aquarius* was equipped with lithium hydroxide canisters to scrub the air, but only for the duration of the lunar landing. Incredibly, those in *Odyssey* were of a different shape and so the experts at Mission Control set about the task of fitting these square pegs into round holes using only the limited items

available to the crew. The solution was a jury-rigged device consisting of plastic hoses, the cover from a mission folder and some electrical tape. Inelegant maybe, life-saving definitely.

When the time came for Swigert to begin the delicate operation of powering up the CM's systems ready for re-entry, he was concerned that the electrical components might not function as they were sodden with condensation. Ironically it was the modifications made to the wiring following the Apollo 1 fire that probably saved the crew of Apollo 13.

Five hours before splashdown Lovell made a manual burn of *Aquarius*'s thruster for a final course correction, and thirty minutes later the SM was jettisoned giving the crew their first glimpse of the damage. 'There's one whole side of that spacecraft missing,' confirmed Lovell. Next to go was *Aquarius*, which was destined to burn up in the Earth's atmosphere. 'Farewell, *Aquarius*, and we thank you,' radioed Mission Control. To which Lovell wistfully added, 'She was a good ship.'

Odyssey splashed down within sight of the recovery vessel USS *Iwo Jima* on 17 April 1970, and the world breathed a collective sigh of relief. Apollo 13 may have been a failure, but it had been a magnificent one.

APOLLO 14 – A VERY TOUGH PLACE

Having launched on 31 January 1971, Apollo 14 was heading for Fra Mauro – the site where unlucky 13 should have landed. In command was Alan Shepard, the only Mercury 7 astronaut to get to the Moon. He had been grounded after becoming the first American in space in 1961, because of a medical condition known as Meniere's disease which affects balance. Put in charge of the astronaut

Alan Shepard finally gets his day as he leads the crew of Apollo 14 to the crew transfer bus which will take them to the launch pad.

corps, he controversially appointed himself as commander of Apollo 14 the moment the doctors gave him the all-clear. What with Shepard's fifteen minutes of space time the cynics were quick to label Apollo 14 as an all-rookie mission. Flying with him were newcomers Stuart Roosa, CM pilot, and Edgar Mitchell, LM pilot.

Trouble first struck on the way to the Moon when Roosa manoeuvred the CM, *Kitty Hawk*, to dock with the LM, *Antares*. On the first attempt he had difficulty achieving

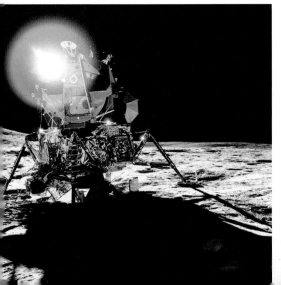

◄◄
The intense sunlight casts a halo around Antares at the Fra Mauro landing site, the intended destination for Apollo 13.

◄
It's one of those obligatory Moon moments – Alan Shepard poses with the Stars and Stripes.

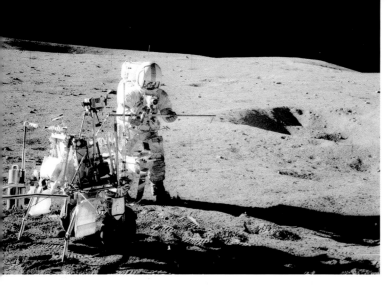

Apollo 14's Al Shepard beside the Modular Equipment Transporter (MET), a glorified golf trolley to transport equipment and samples.

'And when you have tasted flight, you will forever walk the earth with your eyes turned skyward, for there you have been, and there you will always long to return.'

Attributed to Leonardo da Vinci

capture when the latches failed to click shut. The mission was suddenly in deep trouble. Failure to dock would inevitably mean a cancellation of the landing and, after Apollo 13, most likely the end of the programme. He fired the thrusters to try again, acheiving perfect alignment, but with no green light to confirm capture. Over the following hour he made three more attempts, each time eating into the precious fuel reserves. Shepard was all for suiting-up and going outside to fix it, but Mission Control vetoed this plan for being far too dangerous. Roosa was instructed to try one more time, coming in faster to drive *Kitty Hawk* hard against *Antares*, and, thankfully, they had capture.

The next problem reared its ugly head shortly after separation when *Antares'* computer registered an abort signal from a faulty switch. The time-honoured solution

second problem occurred when the radar altimeter failed to lock on to the lunar surface, most probably as a result of the software patch. This was solved by cycling the landing radar breaker, and a signal was acquired at 50,000ft (15,250m), just in the nick of time.

When Alan Shepard stepped off the ladder on 5 February, his first words were, 'It's been a long while, but... we're here!' words that spoke as much of his personal journey as of the many problems that had beset the mission. 'Gazing around at the bleak landscape, it certainly is a stark place at Fra Mauro. It's made all the more stark by the fact that the sky is completely dark. This is a very tough place.'

The astronauts set to work on the first of two EVAs, setting up the scientific equipment and television camera,

Did you know?

While the Moon's diameter is almost one-third of the Earth's, its mass is only 1/81 of that of the Earth. This is why its gravitational pull is a puny 17 per cent, or approximately one-sixth.

of a hefty whack on the appropriate panel made the problem go away, for a while, but if it happened again when the descent stage engine was firing an auto-abort would be initiated. NASA's experts came up with a quick fix, a software modification entered via Antares's computer pad. A

Ed Mitchell studies a traverse map during a frustrating search for Cone Crater. Lunar dust can be seen clinging to his spacesuit.

'In my own view, the important achievement of Apollo was a demonstration that humanity is not forever chained to this planet, and our visions go rather further than that, and our opportunities are unlimited.'

Neil Armstrong, 1999

hammering the flag into the ground and gathering samples. Among the items they unloaded was the Modulised Equipment Transporter, or MET, a trolley which looked for all the world like a golf cart. It was designed to carry equipment and ferry back the heavy samples in order to extend the range of the astronauts' activities.

Just as with Apollo 12, they experienced problems with the dust. 'It is very fine, like talcum powder. It clings to everything,' said Shepard. Outwardly businesslike, he paused briefly at one point to take in the spectacle of this alien world with the crescent of Earth hanging above in the void, and he wept silently from pent up emotion, the relief of the landing and the sheer wonderment of the beauty of his home planet.

As they continued their EVA the two men quickly adapted to the one-sixth gravity, gaining confidence with a bouncing gait; as Mitchell confirmed, 'It's easy. Just a little push and you spring right up.' At times they showed signs of working too intensely, breathing hard and grunting at their exertions; something remedied with a slowing of pace, enabling the moonwalk to be extended to four hours and fifty minutes.

On the second day they were scheduled to explore the 1,000ft (300m)-wide rim of Cone Crater, situated less than a mile from *Antares*, but the climb up the slope to the rim proved too demanding. To add to their troubles the heavily loaded MET dug into the soft soil, and just navigating about this featureless undulating expanse was far harder than expected. Reluctantly they returned to *Antares* to load their samples, and as the world looked on Shepard reached into a pocket and withdrew a small metal flange which he attached to a long aluminium handle. 'Houston, you may recognise what I have in my hand… the handle for the contingency sample. It just so happens to have a genuine six-iron on the bottom.' From another pouch came a golf ball and on his second one-handed swing he sent it flying. 'There it goes! Miles and miles and miles.'

APOLLO 15 – HADLEY RILLE

Apollo 15 blasted-off on 26 July 1971, at a time when the Apollo programme was facing severe cutbacks. The remaining three missions needed to place a far greater emphasis on science than the earlier landings which had been predominantly about getting there and returning safely. It was time to go exploring, which would be achieved by increasing the astronauts' mobility and by putting them through an intensive pre-flight grounding in geological practices and theory.

In command of an all-air force crew was David Scott, who previously flew on Apollo 10, joined by newbies Al Worden, CM pilot, and James Irwin, LM pilot. Their destination was Hadley Rille, a snaking canyon in the Mare Imbrium surrounded by the Appennine Mountains. This made it the most challenging landing so far, with

Falcon the most heavily laden LM, and on 30 July Scott radioed back to Earth, 'Okay Houston, the *Falcon* is on the Plain at Hadley.'

Apollo 15 was to stay on the lunar surface for sixty-seven hours – more than Apollo 11, 12 and 14 put together. The first EVA would take place on the second day after they had rested, but before they settled down Scott poked his head and

➤

Dave Scott, Al Worden and Jim Irwin of the Apollo 15 mission.

➤➤

Tha canyon at Hadley Rille, the landing site for Apollo 15, can be seen snaking across the lower left of this photograph taken from orbit.

shoulders up through *Falcon*'s upper hatch to get a feel for their new home. Much of it was familiar from training, but this did not detract from the majesty of the surroundings; the 11,000ft (3,350m)-high Hadley Delta Mountain to the south with the St George Crater, bigger than twenty-seven football fields, gouged into its slopes.

The major innovation for this mission was the Lunar Roving Vehicle (LRV), which had travelled folded up against *Falcon*'s side. Built by Boeing, it had an empty weight of just 460lb (209kg). Each wheel was driven by an electric motor and it could attain speeds of up to 6 or 8mph (10 to 12km/h), but what no one had predicted was how exciting the ride would be as it bounced with every bump in one-sixth gravity. On the first EVA the astronauts took samples at Elbow Crater before continuing to the

base of the Mount Hadley Delta. When they got back to *Falcon*, Irwin deployed the ALSEP while Scott unloaded their samples. The next task, drilling a pair of 10ft (3m) holes for a heat-flow experiment, proved unexpectedly difficult and was postponed until the following day.

Fully suited-up, Scott and Irwin practise driving the lunar rover on a simulated lunar surface.

The second EVA took them to the slopes of Hadley Mountain in the lunar rover. It performed remarkably well and when the moment came to climb out, Scott

Did you know?

The first wheeled vehicle to drive on the Moon was not the Apollo 15 Lunar Rover which landed on 30 July 1971. This was beaten to it by the Soviet Union's Lunokhod, an eight-wheeled radio-controlled explorer which landed aboard the Luna 17 spacecraft on the Sea of Rains on 17 November 1970. This 7.5ft (2.3m)-long robot vehicle eventually travelled 6.5 miles (10.5km), transmitting close-up pictures of the surface as well as analysis of the soil, and remained in operation for 322 days before contact was lost. A second Lunokhod landed in January 1973, and travelled 23 miles (37km).

and Irwin were taken aback by the steep incline which threatened to send the rover sliding backwards. That would have been disastrous, with a long trek back to *Falcon*. But it was here on these precarious slopes that Irwin made the most important find of the mission; a 4.5 million-year-old piece of primordial crust that would become known as the 'Genesis' rock. 'I think we found what we came for,' Scott informed Mission Control. Back at *Falcon*, they completed the heat flow holes but struggled to drill a

Apollo 15 rollout: the Saturn V is taken out of the Vehicle Assembly Building (VAB) and heads towards Pad 39A, riding with its mobile launch tower atop the crawler-transporter.

Jim Irwin salutes the flag beside the lunar module, Falcon, and the lunar rover. Mount Hadley Delta is in the background. The base of the mountain is approximately 3 miles (5km) away.

deep core sample. For a second 'night' the astronauts were able to sleep without their spacesuits, despite their exhaustion and being grimy with dust.

On the final day the drill continued to sap their strength. What no one had appreciated was that the fine lunar soil had compacted over aeons, leaving no space for the drill to penetrate. With a final effort the core sample was freed, but they still had a fight to dismantle it. The astronauts were

'As I stand out here in the wonders of the unknown at Hadley, I sort of realise there's a fundamental truth to our nature. Man must explore...'

David Scott, Commander Apollo 15,
31 July 1971

The Apollo astronauts took a variety of unauthorised items to the Moon. Alan Shepard of Apollo 14 had some golf balls and the head of a six-iron club, Dave Scott of Apollo 15 took a falcon feather, as well as the 'Fallen Astronaut' figurine in tribute to those who had died for the cause of space exploration, and Charlie Duke of Apollo 16 left a family photograph.

Irwin prepares the lunar rover for Apollo 15's first EVA, with Mount Hadley in the background.

▶▶
The final parking space for the Apollo 15 lunar rover, positioned so that Mission Control could remotely operate the TV camera to get the first images of an ascent stage launch.

glad when word came to head to the edge of the Rille, its deep bottom concealed in the impenetrable shadows.

Returning to *Falcon*, Scott continued the science theme with a demonstration of Galileo's theory that objects of different mass fall at the same rate in a vacuum. He released a falcon feather and his geology hammer and they fell side by side to hit the ground simultaneously. 'Nothing like a

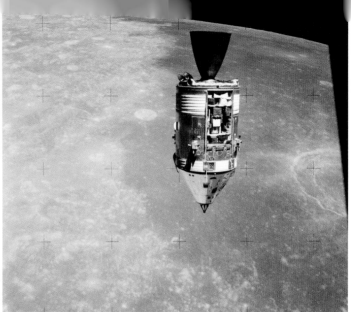

little science on the Moon.' Then, driving alone, Scott took the rover to a small rise about 300ft (91m) away, from where it would transmit pictures of their departure four hours later.

Out of sight, he placed a simple aluminium sculpture of a 'Fallen Astronaut' on the lunar soil, along with a plaque commemorating the astronauts and cosmonauts who had lost their lives in the

The command service module in lunar orbit and displaying the Scientific Instrument Module.

under pressure from NASA, but the crew's reputation was further tarnished when they were formally reprimanded for making a deal to carry unauthorised postal covers to the Moon.

Nonetheless, Apollo 15 had been a scientific triumph. Not only had they returned 170lb (77kg) of material, but while in orbit CM pilot Al Worden had conducted an extensive survey using a new Scientific Instrument Module (SIM) and released a sub-satellite into lunar orbit. On the return flight home he conducted a record-breaking deep space EVA to retrieve film cassettes from the SIM bay.

▲

The unauthorised Fallen Astronaut figure and card which Scott placed on the surface of the Moon to commemorate those who had died in the cause of space exploration.

furtherance of space exploration. Although this had been commissioned secretly from Belgian sculptor Paul Van Hoeydonck, on the understanding that no replicas were to be made, he subsequently advertised them for sale. This offer was withdrawn

In April 1971 John Young made his way back to the Moon. He had been before with Apollo 10, and now he was in command of Apollo 16 for a landing in the Descartes highlands – the highest landing site so far at 18,000ft (5,475m) above lunar 'sea-level'. CM pilot was Tom Mattingly, who had missed out on a place on Apollo 13, luckily or not, and LM pilot was Charles Duke, who served as Capcom when Apollo 11 touched down on the Moon. As with Apollo 15, they were equipped with a lunar rover for their three-day stop-over, including three EVAs and a quest to bag some volcanic rock for the scientists.

By the fifth day of the mission, on 21 April, Young and Duke were in the LM *Orion* easing away from Mattingly in the CM *Casper*. 'We're sailing free,' *Orion* informed Houston. But seconds before the descent was about to begin Mattingly encountered a problem with oscillations in the motors which angled the CSM's main engine nozzle. It was serious enough to automatically put a hold on proceedings; they only had ten hours in which to fix it before supplies were depleted sufficiently

The Apollo 16 crew: Tom Mattingly, who had been stood down from Apollo 13, with John Young and Charlie Duke.

Did you know?

A load of rubbish: it is estimated that mankind has deposited 37,480lb (17,000kg) of assorted debris on the Moon, while only 842lb (382kg) of samples have been returned to Earth by the Apollo and unmanned lunar missions.

for the landing to be aborted. For nearly three hours the two spacecraft continued their close dance until they were given the go-ahead.

'Well, old *Orion* is finally here, Houston,' announced Duke as they touched down nearly six hours late, metres away from a large crater. The next morning John Young backed down the ladder to become the ninth man to walk on the Moon. Duke followed, his enthusiasm slightly dampened by a leak of orange juice inside his helmet. For the next four hours they set about their tasks, installing the TV camera, placing the flag, setting up the ALSEP and drilling holes for the heat sensors. Then it was time to head off in the rover 1 mile (1.6km) west to a series of small craters, including Flag Crater. 'There's just craters on top of craters,' commented Young. It was a chance for

'I was sure we'd go into space; sure we'd go to the Moon and planets; but I didn't really believe I'd live to see it. Or live to see it finished! That's something I never would have dreamed of: that we would go to the Moon, and abandon it after five years!'

Arthur C. Clarke, 1993

him to put the rover through its paces; the bucking-bronco ride churning up a 'rooster tail' of dust in his wake. Once they were safely back in *Orion* they relaxed a little, chatting about their day and unwittingly treating the world to some candid comments on their digestive problems resulting from an overload of orange juice.

A jubilant John Young leaps high into the air as he salutes the flag on Descartes. Stone Mountain dominates the background.

The lunar module, Orion, is seen in the distance in this view from the rover.

Charlie Duke beside the 120ft (36.5m)-wide Plum Crater during the first moonwalk.

Did you know?

The only astronaut to fly Mercury, Gemini and Apollo missions was Wally Schirra who flew on Mercury MR-8, Gemini 6A and Apollo 7. The only one on Gemini, Apollo and the Space Shuttle was John Young who, in addition to Gemini 3 and 10, Apollo 10 and 16, was commander on the first Shuttle launch, STS-1, in 1981, and also STS-9 in 1986.

Duke beside the rover, near Stone Mountain on the second EVA at Descartes.

once Young had cause to reign in his driver for the day, 'Charlie, quit pushing this thing around.' At Stone Mountain they took readings and collected 82lb (37kg) of samples, but there was little evidence of lava flows. After a cautious drive uphill they parked beside a 50ft (15m) crater where they found a crystalline white rock. No Genesis Rock moment, but they methodically gathered the evidence geologists back home wanted.

On day three the flight managers trimmed the EVA to only five hours in order to meet the lift-off time, leaving Young and Duke just long enough to reach North Ray Crater in search of deep bedrock. North Ray is huge, about the size of Arizona's meteor crater, and the impact had scattered debris far and wide. 'Oh, spectacular! Just spectacular!' Young

The following morning the target for the second EVA was Stone Mountain, a rounded hill 2.5 miles (4km) to the south. It was rough going and more than

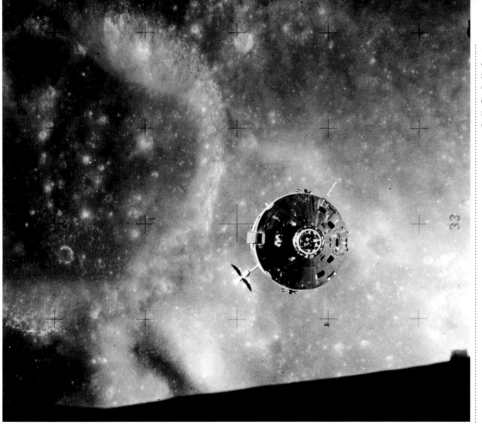

After three days on the surface, John Young and Charlie Duke close in on Thomas Mattingly in the command module, Casper.

reported when they reached it. While taking photographs of the crater they saw a big, dark bolder in the distance and they headed for it. 'Look at the size of that rock!' Duke exclaimed. Being as big as a building, they immediately christened it 'House Rock'.

Apollo 16 had broken the record for time spent on the surface of the Moon, and, as *Orion* hurtled into lunar orbit, one of the astronauts exclaimed, 'What a ride, what a ride!' The homeward leg was straightforward enough, with time for one last television broadcast. Young held a grubby hand up to the cameras. 'Can you see the dirt underneath those fingernails? That's Moon dust.' Then it was Mattingly's turn to go outside to make his way to the SIM to collect the film canisters. Puzzled that he couldn't see any stars, he cracked open his gold-coated visor momentarily, and there they were.

At 12.33 a.m. EST on 7 December 1972 the five engines of a Saturn V's first stage burst into a supernova of light, flooding the darkness and sending Apollo 17 riding a thunderous trail away from Earth. This was the first night-time launch and the final Apollo mission to the Moon.

On board were Commander Gene Cernan, returning three years after Apollo 10, LM pilot Harrison 'Jack' Schmitt and CM pilot Ronald Evans. Originally Joe Engle had been selected as LM pilot, but once it had become clear that this would be the last lunar flight NASA was under pressure to send a scientist and professional geologist, so Schmitt was brought forward from the cancelled Apollo 18. Their destination was the south-eastern rim of the Mare Serenitatis, with a landing between three high massifs in a deep valley known as Taurus-Littrow. The site was selected because it featured boulders along the bases of the mountains which could provide bedrock samples, as well as dark craters which might contain evidence of lava flow.

7 December 1972: Apollo 17 blasts-off on the first night launch of the Saturn V - the final Apollo mission to the Moon.

Professional geologist Harrison 'Jack' Schmitt poses beside Ron Evans and, seated in a rover mock-up, mission commander Gene Cernan.

with feigned mockery. His focus was on the incredible landscape they were exploring.

Before they headed out on the lunar rover to their first destination at the edge of Steno Crater, to the south, there was several hours worth of work to do laying out experiments and wrestling with drilling more core samples. Despite the demanding physical exertion eating into their oxygen reserves, they remained in good humour, with Schmitt bounding across the surface and singing, 'I was strolling on the Moon one day...' Cernan picked up the refrain, 'In the merry, merry month of... December!' Also, with the benefit of their predecessors' experiences, the two men were confident in their movements, although Schmitt in particular often stumbled.

While Evans circled in the CM *America*, Cernan and Schmitt took the LM *Challenger* down to the surface. The first EVA began just four hours later, and as they set about their tasks, Cernan called over to his colleague to look up at Earth. 'Aaah! You seen one Earth, you've seen them all,' Schmitt replied

Day two saw a far more ambitious EVA, and the astronauts headed off on a 5-mile

'It's like trying to describe what you feel like standing on the rim of the Grand Canyon or remembering your first love or the birth of your first child. You have to be there to know what it's like.'

Harrison 'Jack' Schmitt, Apollo 17

(8km) drive to the slopes of the 7,500ft (2,286m)-high South Massif. At times *Challenger* was out of sight, but they only ever travelled to the extent of their ability to walk back in the event of something going wrong with the rover. Near the rim of the 360ft (110m) Shorty Crater, Schmitt's attention was caught by a large fractured boulder. As he was preparing to take some panoramic shots of the scene he noticed a patch of colour at his feet. 'Oh hey, there is orange soil!' The scientists back on Earth thought the astronauts had found what they had been looking for: evidence of volcanic action.

Day three followed with an EVA lasting seven hours and fifteen minutes, taking them north-east to a dark boulder perched on the steep slopes of the North Massif.

Gene Cernan salutes the flag on the Taurus-Littrow region of the Moon.

95

Harrison Schmitt takes samples from beside the split bolder on the third EVA.

Did you know?
In 1990 Japan became the third nation to send a spacecraft to the Moon with its unmanned Hiten orbiter, although this later crashed onto the surface. Since then both the European Space Agency and India have sent their own robot landers.

97

◄
Cernan stands beside the lunar rover. The camera at the front was remotely controlled by Mission Control and provided live colour TV images from the Moon.

▲
When a rear fender was damaged on the rover a repair was improvised using one of the maps and some tape.

Covered in dust, Schmitt uses an adjustable sampling scoop during the second EVA. The rod mounted on a tripod is used to provide a vertical reference and a scale for size. Although everything look grey, this is a colour image.

The ascent stage of Challenger shortly before docking with the command module, America. Afterwards, the ascent stage was crashed into the Moon to provide data for the seismometers left on the surface.

Weary but satisfied with their work; Cernan and Schmitt had spent just over three days on the Moon.

Did you know?
It takes the Moon around 29.5 days to pass through its phases. This is known as the synodic period, and it means that a lunar day is 14.25 Earth days long. It takes 27.3 days for the Moon to orbit the Earth, and this makes it possible to occasionally have two full moons within a single month. When this occurs the second is known as a 'blue moon'.

180,000 miles (290,000km) from Earth, Ron Evans goes on a deep space walkabout to retrieve film cannisters from the SIM bay located on the service module.

A precision splashdown and recovery for the crew of Apollo 17, with USS Ticonderoga in the background.

Finally it was time to go, and, with Schmitt back in *Challenger*, Cernan became the last man to stand on the barren surface of the Moon. 'We leave as we came and, God willing, as we shall return, with peace and hope for all mankind...' Apollo 17 had been the longest lunar mission, spending the most time on the Moon at seventy-five hours, and had covered nearly 21 miles (34km). In total the six Apollo landings returned around 850lb (385kg) of lunar rock, thousands of photographs and a mountain of data collected on the surface and from lunar orbit. However, they came back with more than just rocks; perhaps Apollo's most enduring icon remains a single photograph taken by Ron Evans. Known as the 'blue marble photo', it shows Earth as seen from space. They had gone to explore the Moon, but what they discovered was the importance, the beauty and the fragility of our home planet.

The Apollo Application Programme (AAP) was instigated in 1968 to develop science-based missions using surplus Apollo hardware. AAP had its origins in a number of earlier proposals, including ambitious plans for manned lunar bases and an Earth-orbiting space station. Clearly NASA still wanted to go to the Moon, but financial constraints and political indifference was forcing a re-evaluation of the space

programme. They had to start delivering value-for-money projects with tangible benefits for the tax-payers, so with this in mind the focus shifted to the space station. Several strands from different projects were woven together into what became Skylab, and, with the cancellation of Apollo 18, 19 and 20, the necessary rockets had become available.

Skylab was launched on 14 May 1973 as the third stage of a Saturn V-INT, fitted-out as a two-level workshop/living space, equipped with a multiple docking adapter/airlock module and the Apollo Telescope Mount (ATM) solar observatory. Unfortunately during the launch its micrometeroid thermal shield was torn away and debris from this jammed one of the main solar panel arrays leaving the space station overheated and starved of

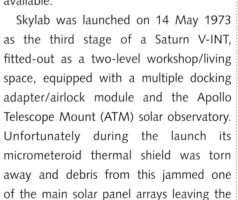

◄

The patched-up Skylab with only one main solar panel array extended, photographed from the departing Skylab 4 crew during a final fly-around.

103

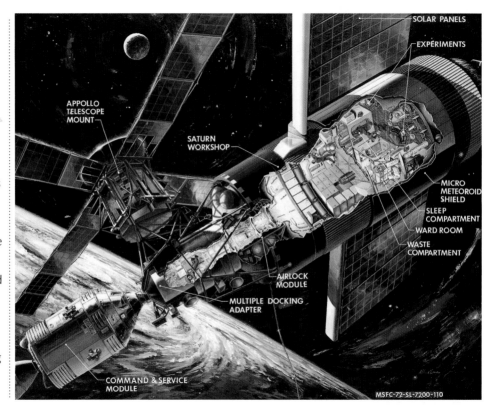

This cutaway diagram of Skylab reveals how spacious NASA's first space station was.

Did you know?

The Apollo-Soyuz Test Project was known by the Soviets as the Soyuz-Apollo Test Project. When it came to designing the docking mechanism neither nation wanted the role of 'female' drogue to the other's 'male' probe, and an androgynous docking apparatus was built.

SOLAR PANELS

EXPERIMENTS

APOLLO TELESCOPE MOUNT

SATURN WORKSHOP

MICRO METEOROID SHIELD

SLEEP COMPARTMENT

WARD ROOM

WASTE COMPARTMENT

AIRLOCK MODULE

MULTIPLE DOCKING ADAPTER

COMMAND & SERVICE MODULE

MSFC-72-SL-7200-110

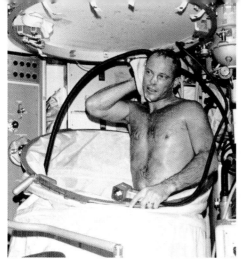

comfort, with dedicated sleeping quarters, an eating area with a window on the world, hot food and other facilities including a shower.

When Skylab was abandoned in February 1974 it was with the prospect that the Space Shuttle would be on the scene in time to kick it into a higher orbit. In the

power. When the first crew headed-off in an Apollo CSM, atop a Saturn 1B, eleven days later, the immediate task was to make emergency repairs. A total of three crews visited Skylab, with the third establishing a new endurance record of eighty-four days in orbit. Compared with the lunar astronauts they enjoyed a high degree of

◄

All mod cons – Jack Lousma relaxes in a warm shower in Skylab's crew quarters. The shower curtain would be drawn right up to the ceiling and afterwards loose water was extracted by a vacuum pump.

◄

Owen Garriot performs an EVA during the Skylab 3 mission. He is seen beside the Apollo Telescope Mount (ATM) deploying a particle collection experiment.

Visualisation of the Apollo spacecraft carrying the docking adaptor, facing the Soviet Soyuz 19.

event the Shuttle didn't fly until 1981 and, as there were no alternative spacecraft available in the interim, Skylab was left to burn up in the Earth's atmosphere in 1979.

Despite Skylab's ignominious fall from grace, the Apollo programme had one last clarion call; the Apollo-Soyuz Test Project (ASTP) which brought together the two former space rivals. Aboard an Apollo CSM were Commander Tom Stafford, Deke Slayton – one of the Mercury 7,

who finally got his space wings following surgery to correct a heart murmur – and Vance Brand. In the Soviet Soyuz 19 were Aleksei Leonov, the first man to walk in space, and Valeri Kubasov. On the evening of 17 July 1975 the two spacecraft linked up via a special docking module carried by the Apollo, the crews shaking hands and exchanging greetings in a symbolic union that would lead to future joint efforts in space, culminating with the International Space Station (ISS).

After Apollo came the Space Shuttle, which promised an era of reliable launches into

'You carry with you your own body-orientated world... in which up is over your head, down is below your feet... and you take this world around with you wherever you go.'
Joe Kerwin on his experiences with Skylab 2

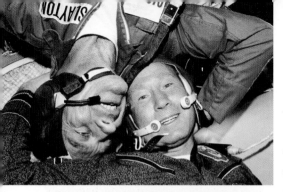

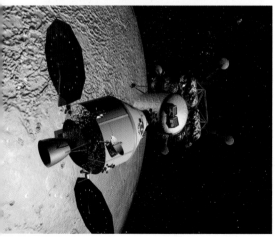

Earth orbit, achieved with a near businesslike regularity for almost thirty years and making headlines only when things went disastrously wrong. However, the Shuttle did little to ignite passion in the hearts of the next generation of space sailors in the way that the Moon shots had. It's a big space bus with precious little to do with exploration; frankly, nobody was 'boldly going' anywhere. It wasn't even the cheap Space Transportation System that it had been heralded as, and, consequently, the Shuttle is slated for retirement in 2010, although there is talk of keeping it flying until a successor, Project Constellation, is ready.

Project Constellation consists of an Apollo-style spacecraft known as *Orion* carrying a

'We'll go into orbit. We'll go to the Moon. This business has no limits.'

Sir Richard Branson, Virgin Galactic, 2005

◄

Deke Slayton of Apollo comes head-to-head with cosmonaut Aleksei Leonov of Soyuz 19 following the Apollo-Soyuz rendezvous in July 1975. This was the final flight of an Apollo spacecraft.

◄

Visualisation of an Orion CSM docked with an Altair lunar lander. All being well NASA say they will be back on the Moon by 2019, possibly in time for the fiftieth anniversay of the Apollo 11 landing.

The Orion space capsule is rolled out at NASA's Langley Research Center in Virginia, prior to unmanned suborbital test flights at the White Sands Missile Range in New Mexico.

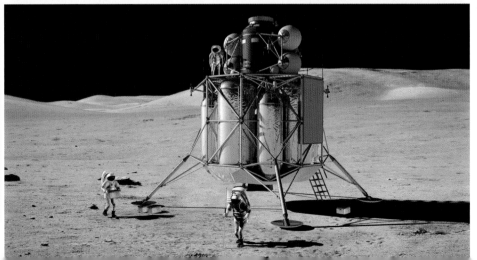

Visualisation of the Altair lander. As with the Apollo lunar module it is designed to fly only in the vacuum of space and accordingly aerodynamic streamlining is not an issue.

crew of up to six, launching on the Ares I rocket, thus utilising existing Shuttle booster technology. It looks for all the world like a beefed-up Apollo, as NASA Administrator Michael Griffin freely admitted in 2005. 'Think of it as Apollo on steroids. If much of it looks the same, that's because the fundamental laws of physics haven't changed recently.' By which he means that the conical capsule is still the best shape for re-entering the atmosphere. The capsule, or Crew Exploration Vehicle (CEV), has two-and-a-half-times the volume of the Apollo capsule, with a 'glass cockpit' instrument layout, an 'autodock' low-impact docking system and vastly improved computers. It will also be a reusable craft, capable of making up to ten flights.

Attached to the CEV is a cylindrical service module housing the propulsion systems and expendable supplies, as well as carrying solar panels to eliminate the need for fuel cells. The first manned flight is scheduled for no later than 2014. Constellation is envisaged as a pragmatic 'pay-as-you-go' solution to meet NASA's future aspirations, including not only a return to the Moon from 2019 onwards, but also to Mars and beyond. Accompanying *Orion* to the Moon will be the Altair Lunar Surface Access Module (LSAM), a direct descendant of the Apollo LM.

Where the Apollo programme showed the way, others will follow.

'Space travel is bunk.'

Sir Harold Spencer Jones, Astronomer Royal, 1957 (two weeks before Sputnik 1)

Did you know?
The US Commercial Space Launch Amendments Act of 2004 put the regulatory framework in place for commercial space flight. The law established an 'informed consent' regime for carrying passengers, and created a new experimental launch permit for the testing and development of reusable suborbital launch vehicles.

MANNED APOLLO FLIGHTS

Mission	Crew	Launch vehicle	Lift-off	Splashdown	Mission summary
Apollo 1 AS-204	Gus Grissom, Roger Chaffee, Ed White	Saturn 1B AS-204	Intended as the first manned flight of Apollo, the three astronauts died in a capsule fire on the launch pad, 27 Jan 1967.		
Apollo 7	Walter Schirra, Walter Cunningham, Don Eisele	Saturn 1B AS-205	11 Oct 1968	22 Oct 1968	Testing Apollo CSM spacecraft in low Earth orbit.
Apollo 8	Frank Borman, William Anders, James Lovell	Saturn V AS-503	21 Dec 1968	27 Dec 1968	First flight around the moon.
Apollo 9	Jim McDivitt, Rusty Shweickart, David Scott	Saturn V AS-504	3 Mar 1969	13 Mar 1969	First flight of the LEM, tested in low Earth orbit. CM: *Gumdrop*. LM: *Spider*.
Apollo 10	Thomas Stafford, Gene Cernan, John Young	Saturn V AS-505	18 May 1969	26 May 1969	Testing LEM in lunar orbit. CM: *Charlie Brown*. LM: *Snoopy*.
Apollo 11	Neil Armstrong,* Buzz Aldrin,* Michael Collins	Saturn V AS-506	16 Jul 1969	24 Jul 1969	First Moon landing, Sea of Tranquility. CM: *Columbia*. LM: *Eagle*.

* The twelve astronauts who walked on the Moon.

Mission	Crew	Launch vehicle	Lift-off	Splashdown	Mission summary
Apollo 12	Pete Conrad,* Alan Bean,* Richard Gordon	Saturn V AS-507	14 Nov 1969	24 Nov 1969	Landing at the Ocean of Storms, visited Surveyor 3. CM: *Yankee Clipper*. LM: *Intrepid*.
Apollo 13	James Lovell, Fred Haise, John Swigert	Saturn V AS-508	11 Apr 1970	17 April 1970	Explosion on CSM forces cancellation of Moon landing. CM: *Odyssey*. LM: *Aquarius*.
Apollo 14	Alan Shepard,* Stuart Roosa,* Edgar Mitchell	Saturn V AS-509	31 Jan 1971	9 Feb 1971	Landing at Fra Mauro region. CM: *Kitty Hawk*. LM: *Antares*.
Apollo 15	David Scott,* James Irwin,* Alfred Worden	Saturn V AS-510	26 Jul 1971	7 Aug 1971	Explored Hadley Rille region, first use of the lunar rover. CM: *Endeavour*. LM: *Falcon*.
Apollo 16	John Young,* Charlie Duke,* Thomas Mattingly	Saturn V AS-511	16 Apr 1972	27 Apr 1972	Landing in Descartes highlands region, second mission with rover. CM: *Casper*. LM: *Orion*.
Apollo 17	Gene Cernan,* Harrison Schmitt,* Ronald Evans	Saturn V AS-512	7 Dec 1972	19 Dec 1972	Final Moon landing, explored Taurus-Littrow Valley with rover. CM: *America*. LM: *Challenger*.

Mission	Crew	Launch vehicle	Lift-off	Splashdown	Mission summary
Skylab	unmanned	Saturn V AS-513 INT-21	14 May 1973	11 Jul 1979	NASA's first space station, remained in orbit until July 1979. Last flight of a Saturn V; the INT-21 was a two-stage version.
Skylab SL-2	Pete Conrad, Joe Kerwin, Paul Weitz	Saturn 1B AS-206	25 May 1973	22 Jun 1973	First of three missions to Skylab, with emphasis on making repairs to solar panel array. Set new endurance record of twenty-eight days in space.
Skylab SL-3	Alan Bean, Owen Garriot, Jack Lousma	Saturn 1B AS-207	28 Jul 1973	25 Sep 1973	Fifty-nine-day mission.
Skylab SL-4	Jerry Carr, Ed Gibson, Bill Pogue	Saturn 1B AS-208	16 Nov 1973	8 Feb 1974	Over eighty-four days living in space, final Skylab mission.
ASTP	Thomas Stafford, Vance Brand, Deke Slayton	Saturn 1B AS-209	15 Jul 1975	24 Jul 1975	Apollo-Soyuz Test Project, rendezvous with Soviet Soyuz spacecraft manned by Aleksei Leonov and Valeri Kubasov. Final Apollo flight and last flight of Saturn 1B.

WHAT'S LEFT OF APOLLO?

There is a surprisingly wide array of Apollo hardware on display. The best items are to be found on the Moon with six landing sites, complete with LM descent stages and three Lunar Rovers, all with low mileage. However, if getting there is a little out of your budget then you could start a little closer to home.

Command service modules:
With the flown missions, the only part of the Apollo spacecraft to return to Earth was the CM capsule:

Apollo 1	In storage at Langley Research Centre, not on display
Apollo 7	Frontiers of Flight Museum, Love Field, Texas
Apollo 8	Museum of Science and Industry, Chicago
Apollo 9	San Diego Aerospace Museum
Apollo 10	Science Museum, London
Apollo 11	National Air and Space Museum, Washington DC
Apollo 12	Virginia Air and Space Museum
Apollo 13	Kansas Cosmosphere and Space Center
Apollo 14	United States Astronaut Hall of Fame, Florida

After assembly in the 525ft (160m)-tall Vehicle Assembly Building (VAB), an Apollo Saturn V is carried on the crawler-transporter the short distance to the launch pad.

The Apollo 8 capsule on board the recovery ship, USS Yorktown. The spherical inflation bags are designed to right the capsule should it invert in the water.

Apollo 15	US Air Force Museum, Wright-Patterson, Ohio
Apollo 16	US Space and Rocket Center, Houston, Texas
Apollo 17	Johnson Space Center, Houston, Texas
ASTP	Kennedy Space Center, Florida
Skylab 2	National Museum of Naval Aviation, Pensecola, Florida
Skylab 3	John H. Glenn Research Center, Cleveland, Ohio
Skylab 4	National Air and Space Museum, Washington DC

In addition there are a number of backup or non-flight test vehicles, displayed at several locations:

CSM-001	Cradle of Aviation, Long Island, NY
CSM-007	Museum of Flight, Seattle, Washington
CSM-009	Strategic Air and Space Museum, Ashland, Nebraska
CSM-010	US Space and Rocket Center, Huntsville, Alabama
CSM-011	National Air and Space Museum, Washington DC
CSM-020	Fernbank Science Center, Atlanta
CSM-098	Academy of Science Museum, Moscow
CSM-119	Kennedy Space Center, Florida

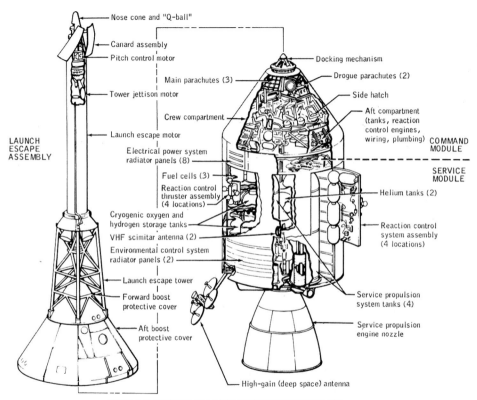

Nose cone and "Q-ball"

Canard assembly

Pitch control motor

Docking mechanism

Drogue parachutes (2)

Main parachutes (3)

Tower jettison motor

Side hatch

Crew compartment

Aft compartment
(tanks, reaction
control engines,
wiring, plumbing)

LAUNCH
ESCAPE
ASSEMBLY

COMMAND
MODULE

Launch escape motor

SERVICE
MODULE

Electrical power system
radiator panels (8)

Fuel cells (3)

Reaction control
thruster assembly
(4 locations)

Helium tanks (2)

Cryogenic oxygen and
hydrogen storage tanks

VHF scimitar antenna (2)

Reaction control
system assembly
(4 locations)

Environmental control system
radiator panels (2)

Launch escape tower

Forward boost
protective cover

Service propulsion
system tanks (4)

Aft boost
protective cover

Service propulsion
engine nozzle

High-gain (deep space) antenna

**APOLLO COMMAND AND SERVICE MODULES
AND LAUNCH ESCAPE SYSTEM**

116

Lunar modules:

As the LM was never designed to return to Earth,* the following are either test or un-flown examples:

LM-2   National Air and Space Museum, Washington DC
LM-9   Kennedy Space Center, Florida
LM-13   Cradle of Aviation, Long Island, New York

*The ascent stage of Apollo 10's *Snoopy* remains in heliocentric orbit around the Sun, the sole survivor of the flown LMs.

Saturn 1B launch vehicles:
SA-209   Kennedy Space Center, Florida
SA-211   First stage at 1-65 Alabama Welcome Center, second stage at US Space and Rocket Center, Huntsville, Alabama

Saturn V launch vehicles:
SA-500F   US Space and Rocket Center, Huntsville, Alabama
SA-513   Third stage at Johnson Space Center, Texas

LUNAR MODULE ASCENT STAGE

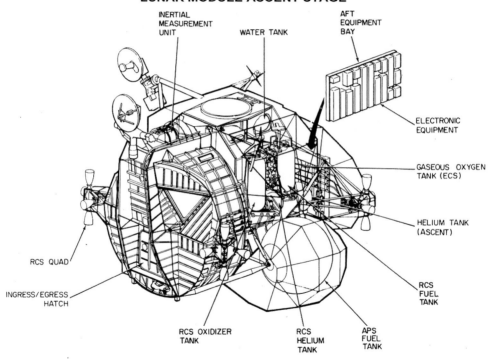

INERTIAL MEASUREMENT UNIT

WATER TANK

AFT EQUIPMENT BAY

ELECTRONIC EQUIPMENT

GASEOUS OXYGEN TANK (ECS)

HELIUM TANK (ASCENT)

RCS FUEL TANK

RCS QUAD

INGRESS/EGRESS HATCH

RCS OXIDIZER TANK

RCS HELIUM TANK

APS FUEL TANK

118

LUNAR MODULE DESCENT STAGE

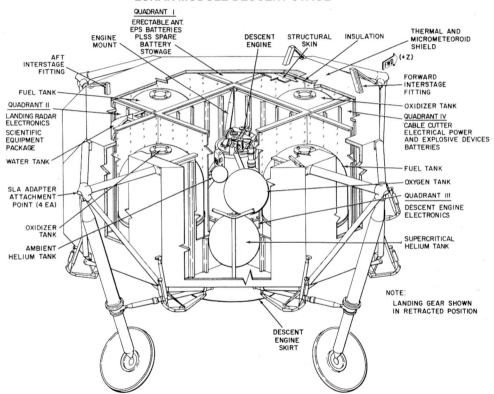

QUADRANT I
ERECTABLE ANT.
EPS BATTERIES
PLSS SPARE
BATTERY
STOWAGE

ENGINE
MOUNT

DESCENT
ENGINE

STRUCTURAL
SKIN

INSULATION

THERMAL AND
MICROMETEOROID
SHIELD

AFT
INTERSTAGE
FITTING

FWD (+Z)

FORWARD
INTERSTAGE
FITTING

FUEL TANK

OXIDIZER TANK

QUADRANT II

LANDING RADAR
ELECTRONICS

SCIENTIFIC
EQUIPMENT
PACKAGE

WATER TANK

QUADRANT IV
CABLE CUTTER
ELECTRICAL POWER
AND EXPLOSIVE DEVICES
BATTERIES

FUEL TANK

OXYGEN TANK

SLA ADAPTER
ATTACHMENT
POINT (4 EA)

QUADRANT III

DESCENT ENGINE
ELECTRONICS

OXIDIZER
TANK

AMBIENT
HELIUM TANK

SUPERCRITICAL
HELIUM TANK

NOTE:
LANDING GEAR SHOWN
IN RETRACTED POSITION

DESCENT
ENGINE
SKIRT

119

| SA-514 | First stage at Johnson Space Center, Texas, second and third stages at Kennedy Space Center, Florida |
| SA-515 | First stage at Michoud Assembly Facility, New Orleans, second stage at Johnson Space Center, third stage converted to backup Skylab workshop at National Air and Space Museum, Washington DC. |

Apollo astronauts:

At the time of writing (January 2009), forty years on from the Moon programme, the following Apollo astronauts are no longer with us:

Gus Grissom, Ed White and Roger Chaffee of Apollo 1; Wally Schirra and Don Eisele of Apollo 7; Pete Conrad of Apollo 12 and Skylab 2; John Swigert of Apollo 13; Alan Shepard and Stuart Roosa of Apollo 14; Jim Irwin of Apollo 15; Deke Slayton of the ASTP.

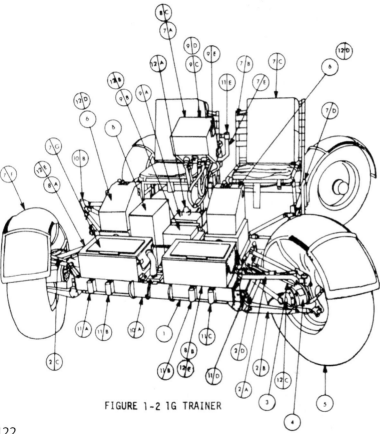

FIGURE 1-2 1G TRAINER

1. CHASSIS

2. SUSPENSION SYSTEM

   A. UPPER ARM
   B. LOWER ARM
   C. DAMPER
   D. TORSION BAR

3. STEERING SYSTEM (FORWARD AND REAR)

4. TRACTION DRIVE

5. WHEEL

6. DRIVE CONTROLLERS

7. CREW STATION

   A. CONTROL AND DISPLAY CONSOLE
   B. SEAT
   C. REMOVEABLE PAD (FOR UNSUITED CREW USE)
   D. OUTBOARD HANDHOLD
   E. INBOARD HANDHOLD
   F. FENDER
   G. SIMULATED DUST COVER

8. POWER SYSTEM

   A. BATTERY #1
   B. BATTERY #2
   C. INSTRUMENTATION

9. NAVIGATION

   A. DIRECTIONAL GYRO UNIT (DGU)
   B. SIGNAL PROCESSING UNIT (SPU)
   C. INTEGRATED POSITION INDICATOR (IPI)
   D. SUN SHADOW DEVICE
   E. ATTITUDE INDICATOR

10. DEPLOYMENT SIMULATION

    A. FORWARD CHASSIS SADDLE SIMULATOR
    B. TRIPOD SIMULATORS (BOTH SIDES)

11. PAYLOAD INTERFACE

    A. TV CAMERA RECEPTACLE
    B. LCRU RECEPTACLE
    C. HIGH GAIN ANTENNA RECEPTACLE
    D. AUXILIARY CONNECTOR
    E. LOW GAIN ANTENNA RECEPTACLE

12. THERMAL CONTROL

    A. DGU HEAT EXCHANGER
    B. SPU HEAT EXCHANGER
    C. TRACTION DRIVE BLOWERS (4)
    D. DCE BLOWERS
    E. BATTERY BLOWER

◄

Gene Cernan and Harrison Schmitt of Apollo 17 preparing the lunar rover for a pre-flight mission simulation, with the lunar module in the background.

123

A. COMMAND & SERVICE MODULE

1. SPS Engine
2. Running Lights (8 places)
3. Scimitar Antenna
4. Docking Light
5. Pitch Control Engines
6. Crew Hatch
7. Pitch Control Engines
8. Rendezvous Window
9. EVA Handholds
10. EVA Light
11. Side Window
12. Roll Engines (2 places)
13. EPS Radiator Panels
14. SM RCS Module (4 places)
15. ECS Radiator

B. MULTIPLE DOCKING ADAPTER

1. Axial Docking Port Access Hatch
2. Docking Target
3. Exothermic Experiment
4. Infrared Spectrometer Viewfinder
5. Atmosphere Interchange Duct
6. Area Fan
7. Window Cover
8. Cable Trays
9. Inverter Lighting Control Assembly
10. L-Band Antenna
11. Proton Spectrometer
12. Running Lights (4 places)
13. Infrared Spectrometer
14. Film Vault 4
15. Film Vault 1
16. SO82 (A&B) Canisters
17. M512/M479 Experiment
18. Area Fan
19. Composite Casting
20. Film Vault 2
21. TV Camera Input Station
22. Utility Outlet
23. M168 STS Miscellaneous Stowage Container
24. Redundant Tape Recorder
25. Radial Docking Port
26. 10-Band Multispectral Scanner
27. TV Camera Input Station
28. Temperature Thermostat
29. Radio Noise Burst Monitor
30. ATM C&D Console

C. AIRLOCK MODULE

1. Deployment Assembly Reels and Cables
2. Solar Radio Noise Burst
3. Handrails
4. DO21/DO24 Sample Panels
5. (Removed)
6. Clothesline (EVA use)
7. Permanent Stowage Container
8. STA IVA Station
9. Nitrogen Tanks (6 places)
10. Oxygen Tanks (6 places)
11. Molecular Sieve
12. Condensate Module
13. Electrical Feedthru Cover
14. Electronics Module 1
15. EVA Hatch
16. Airlock Instrumentation Panel
17. Molecular Sieve
18. STS C&D Console
19. ATM Deployment Assembly
20. Battery Module (2 places)
21. EVA Panel
22. Airlock Internal Hatches (2 places)
23. S193 Microwave Scatterometer Antenna
24. Running Lights (4 places)
25. Handrails
26. Stub Antennas (2 places)
27. Thermal Blanket
28. Discone Antenna (2 places)

APOLLO TELESCOPE MOUNT

1. Command Antenna
2. Telemetry Antenna
3. Solar Array Wing 1
4. Solar Array Wing 2
5. Solar Array Wing 3
6. Solar Array Wing 4
7. Command Antenna
8. Telemetry Antenna
9. Sun-End Work Station Foot Restraint
10. Temporary Camera Storage
11. Quartz Crystal Microbalance (2 places)
12. Acquisition Sun Sensor Assembly
13. ATM Solar Shield
14. Clothesline Attach Boom
15. EVA Lights (8 places)
16. Sun-End Film Tree Stowage
17. Handrail
18. SO82-B Experiment Aperture Door
19. Ha-2 Experiment Aperture Door
20. SO82-A Film Retrieval Door
21. SO82-A Experiment Aperture Door
22. SO54 Experiment Aperture Door
23. Fine Sun Sensor Aperture Door
24. SO56 Experiment Aperture Door
25. SO52 Experiment Aperture Door
26. Ha-1 Experiment Aperture Door
27. SO55A Experiment Aperture Door
28. SO82-B2 Experiment Aperture Door
29. SO82-B Film Retrieval Door
30. Canister Solar Shield
31. Canister
32. Canister Radiator
33. Rack
34. Charger-Battery-Regulator Modules (18 places)
35. Handrail
36. CMG Inverter Assembly (3 places)
37. Control Moment Gyro (3 places)
38. Solar Wing Support Structure (3 places)
39. ATM Outriggers (3 places)

E. ORBITAL WORKSHOP

1. OWS Hatch
2. Nonpropulsive Vent Line
3. VCS Mining Chamber and Filter
4. Stowage Ring Containers (24 places)
5. Light Assembly
6. Water Storage Tanks (10 places)
7. TO13 Force Measuring Unit
8. VCS Fan Cluster (3 places)
9. VCS Duct (3 places)
10. Scientific Airlock (2 places)
11. WMC Ventiation Unit
12. Emergency Egress Opening (2 places)
13. M509 Nitrogen Bottle Stowage
14. SO19 Optics Stowage Container
15. S149 Particle Collection Container
16. SO19 Optics Stowage Container
17. Sleep Compartment Privacy Curtains (3 places)
18. M131 Stowage Container
19. VCS Duct Heater (2 places)
20. M131 Rotating Chair Control Console
21. Power and Display Console
22. M131 Rotating Chair
23. WMC Drying Area
24. Trash Disposal Airlock
25. OWS C&D Console
26. Food Freezers (2 places)
27. Food Preparation Table
28. M171 Ergometer
29. MO92 Lower-Body Negative Pressure
30. Stowage Lockers
31. Experiment Support System Panel
32. Biomedical Stowage Cabinet
33. M171 Gas Analyzer
34. Biomedical Stowage Cabinet
35. Meteoroid Shield
36. Nonpropulsive Vent (2 places)
37. TACS Module (2 places)
38. Waste Tank Separation Screens
39. TACS Spheres (22), Pneumatic Sphere
40. Refrigeration System Radiator
41. Acquisition Light (2 places)
42. Solar Array Wing (2 places)

SKYLAB

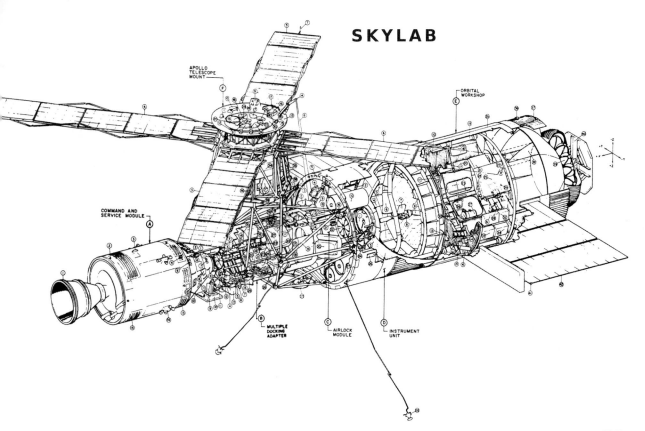

APOLLO
TELESCOPE
MOUNT

ORBITAL
WORKSHOP

COMMAND AND
SERVICE MODULE

MULTIPLE
DOCKING
ADAPTER

AIRLOCK
MODULE

INSTRUMENT
UNIT

APOLLO SPACESUITS

The spacesuit worn by all Apollo astronauts is the A7L which was designed and manufactured by ILC Dover in Delaware. It consists of a one-piece torso-limb suit with joints at the shoulders, elbows, wrists, hips, ankles and knee joints. Covering this is the Integrated Thermal Micrometeroid Garment (ITMG) made up of thirteen layers of material to protect the astronaut from solar radiation and micrometeorites: rubber-coated nylon on the inside, then five layers of aluminised Mylar, four layers of non-woven Dacron, two of aluminised Kapton film/Beta marquisette laminate, and Teflon-coated Beta filament cloth. The joints, made of synthetic rubber, have link-net meshing to prevent the suit from ballooning at these points. The famous fish-bowl helmet and the gloves are joined via metal ring connectors.

The Apollo suits were tailored specifically for each astronaut, although the practice nowadays is to produce a range of standardised sizes. Getting in and out was no easy task, with access from the back where a vertical zipper goes from the shoulder assembly down to the crotch. Beneath the suits, the moonwalkers wore an extra three-layer Liquid Cooling and Ventilation Garment (LCG) which had miles of plastic tubing to cool them with water supplied from the life-support backpack.

The spacesuit was upgraded for the final three lunar flights, Apollo 15, 16 and 17, to allow greater flexibility as well as improved visibility when operating the Lunar Rover. Designated as the A7LB, the new suit incorporated extra joints at the neck and waist, and the backpacks were modified to carry more oxygen, lithium hydroxide, power and cooling water for the longer EVAs.

There were several variations to the standard suit. For example the CM pilot's version was usually without the unnecessary EVA hardware and had a cover layer for fire and abrasion protection. They did not wear the LCG. Skylab's astronauts wore a slightly modified A7LB suit with a less expensive ITMG and visor assembly, and for the ASTP no EVA visors or gloves were required.

Apollo 17 commander, Gene Cernan, wearing the AL7B spacesuit.

GLOSSARY

AAP	Apollo Applications Programme
ASTP	Apollo-Soyuz Test Project
ALSEP	Apollo Lunar Surface Experiment Package
Capcom	Capsule Communications, the person in Mission Control who spoke to the Apollo spacecraft
CEV	Crew Exploration Vehicle
CM	Command Module
CSM	Command and Service Module
EVA	Extra Vehicular Activity
ITMG	Integrated Thermal Micrometeroid Garment
LCG	Liquid Cooling and Ventilation Garment
LEM	Lunar Excursion Module – an earlier designation for the LM
LM	Lunar Module
LOI	Lunar Orbit Insertion
LOR	Lunar Orbit Rendezvous
LRV	Lunar Roving Vehicle, or Lunar Rover
LSR	Lunar Surface Rendezvous
LSAM	Lunar Surface Access Module
MET	Mobile Equipment Transporter – only used on Apollo 14
NASA	National Aeronautics and Space Administration
SIM	Scientific Instrument Module
SM	Service Module
SMU	Self Manoeuvring Unit
SPS	Service Propulsion System – the main engine on the CSM
STS	Space Transportation System – NASA's designation for the Space Shuttle
SWC	Solar Wind Composition experiment
TEI	Trans-Earth Injection
TLI	Trans-Lunar Injection
VAB	Vehicle Assembly Building

Sources

The quotations in this book come from a number of sources including NASA's official mission transcripts and website, media interviews and the following publications:

The Right Stuff by Tom Wolfe; *Schirra's Space* by Wally Schirra with Richard N. Billings; *Countdown* by Frank Borman with Robert J. Serling; *Liftoff* by Michael Collins; *From the Earth to the Moon* by Jules Verne; *A Man on the Moon* by Andrew Chaikin.